Alfred Perceval Graves

Stellar Dust

Alfred Perceval Graves

Stellar Dust

ISBN/EAN: 9783337373535

Printed in Europe, USA, Canada, Australia, Japan

Cover: Foto ©berggeist007 / pixelio.de

More available books at **www.hansebooks.com**

STELLAR DUST

OR THE

LIFE FORCE

An Original Work of Advanced Scientific Ideas
on World-Building and Life-Producing.

The Origin of Man by Spontaneous Generation——Evolution
Explained Away — The Cause of Growth, Use
and Decay—Zodiacal Construction of the
Brain and Body According to
the Law of Twelve.

—BY—

PROF. P. A. GRAVES, ASTROLOGER

SAN FRANCISCO:
1899.

PREFACE.

"Man, that is born of woman, is of few days and full of trouble; he cometh forth as a flower. and is cut down; he fleeth like a shadow, and continueth not."

This volume is the product of many years of careful study and intense mental labor, under the most trying circumstances possible for an author to be placed, being without home, money, sympathy or assistance in the self-imposed task; besides a strong opposition to encounter at almost every step has had the effect to delay its publication.

But financial embarrassment has been the impeeding force and the delaying power, since having been compelled to go forth each day to work for bread, and writing only when not driven to earn food, and that too in the most depressed financial condition ever known to the people of the far west.

These, together with a helpless family of motherless children offered no assistance to the circumstances above mentioned.

In the face of these facts it would be a surprise to the author if the book was more commendable than otherwise, and though it is far inferior to what the author intended it should be, it is nevertheless all the circumstances would permit. It ought to have been through the press three years earlier, but owing to the above named circumstances it has been delayed.

I have tried to enlist the attention of all classes, from the millionaire down to the servant girl of ebon hue, but failed in each and every attempt.

I tried nearly, if not quite all of the printers south of Sacramento to San Diego, California, but always met with the same kind of encouragement, which was: to try some one else, which I did. I even wrote East with no better success; but I made up my mind in the beginning that it took all these mild ingredients to make up the awful dose of disappointments, which has in the past, and still must be taken by all those, who are foolish enough to try to introduce a new idea to the world. I think it is pretty nearly true what a gentleman in San Francisco once remarked concerning new ideas; that is, that it cost 40,000 dollars and required two generations to get a new idea before the American people. In view of this fact I concluded to write for future generations, and let the book bide its time, for money I have none, and $40,000 friends are not hovering conveniently near with stringless sacks and bursting with fullness. The foregoing are all of the apologies I have to offer to the considerate reader.

But some of my sympathizing friends have hinted that they would like to add a word of apology to my own, for no other reason than to explain that I do not rush to the popular trough, drink with the herd and bellow for coin.

But as that would be a very unnatural pleasure for me to enjoy, I must deem all such apologies out of place. Had I consulted public opinion this book would never have been written. And now it is too late to join the popular procession, however pleasant and profitable it might be to myself and others.

I respect public as well as private opinions when they are right, but I entertain no respect for voluntary slavery in religion, henchmen in politics, nor sycophants in science. I have a profound regard for truth, but no respect for falsehood nor deception. I reverence no doctrine, dogma, subject nor science, because some one else did or does. I only admire them for the truth they embrace. I object to wrong whereever I find it, and accept nothing on authority. No man is so good or great, that I fear to criticise his errors, nor so bad that I cannot accept the good he offers. We are only mortals at best, ruled by the same natural laws, differing only in intensity.

> There is no royal name nor blood,
> That men should love or fear,
> But royal deeds above the clouds
> Should make their memory dear.

A physician once remarked that it was a daring thing to do, to question the authority of the old teachers on physiological questions. I replied, that it required no special degree of courage to oppose an error after it has been proven one, that only cowards feared to speak in behalf of truth.

Books are too often written, not to defend truth nor to advance the cause of humanity, but to catch dollars. I test all metals in my crucible and try them as silver is tried. If I am in error on any point in

this profound philosophy, the reader may be doubly
assured that it was not maliciously nor superstitiously
obtained, for I never possessed a pet idea that I
could not release from the cage of my fancy in a
moments warning, if necessary.

I sometimes advance ideas which I cannot support excetp in a logical way, but I make no assertions to-day, that I fear may be overthrown to-morrow, but in case I should, I will thank the one
who does the kindly act.

I am awake to the fact, that world-building
and life-creating is not a safe nor sure business to
engage in during this age. It would do for Moses and
other speculative minds to attempt such hazardous
things in other days when people were imbued with
faith and strangers to the facts of nature.

But in a scientific age like this, when the mental
status is reversed and facts come first, the prospect
for success in gaining public recognition is not so
flattering as formerly. Nevertheless I am going to
launch a theory on the turbid waters of chance, and
try its powers of endurance in weathering the storms
of criticisms from scientific elements.

I beg no points, nor ask for charity; my work is
open to criticism.

If it is too weak to withstand the buffets of the
breakers, it must succumb to their fury, and be dashed
to pieces on the rock bound shore of the sea of science.
The reader will learn, before advancing very far in
the perusal of this work, that the author is not a

Darwinian evolutionist, but a spontaneous productionist, and bases his philosophy on the proposition that man originally was and now is a product of nature's laws, which are executed by the heavenly bodies, and that they created him in his present form, and not through the unknown laws of evolution, nor in the image of his "maker". Moreover, the great laws which created him must necessarily rule, and sustain him from his coming on till his going off of the stage of action.

The author also claims for the same laws the power to create species.

If they can create a single form of life, they can create many forms, since it is only necessary for them to create a nucleus in order to create any form of life. If they could create a single nucleus they could create many more. If more, their power is scarcely limited.

The creation of a nucleus is all the difficulty there is attending the creation of any and all forms of life; and since they can be created with ease and accuracy evolution is unnecessary.

Dissecting the human body, and separating it into its primary parts, as they were put together by the zoadical forces, may be regarded by some as hazardous and uncertain work. But the reader can better judge the merits of the point in question after he has finished the book.

The attempt to disprove the immortality of man will doubtless meet with more opposition than

all other questions involved in the text; for few, if
any, wish to believe that they will not be permitted to
hear the glad notes of Gabriel's sounding horn when
the day arrives for the grave to give up its dead, and
the angry sea to disgorge its many victims, that all
shall "come forth to be judged of the deeds done in
the body", and each assigned to his place on the right
or left as the case may be, there to remain forever.

Of my own choice I would deprive no one of the
innocent pleasure of witnessing his neighbor's sentence
to everlasting punishment by the stern command of
the tenderhearted dispenser of justice.

"Depart ye accursed into the everlasting fire, prepared for the devil and his angels". Nor would I
deprive him of receiving his own reward of everlasting
life for his own good works: "Well done, good and
faithful servant; enter into the joys of thy lord."

But from necessity I must judge the result of the
created from the character of the creator, and not from
a personal opinion as to what should be the result
according to my own views of justice. It matters not
to her what man believes or hopes for; whether creating or destroying, nature executes her laws to the
letter on all occasions. We only deceive ourselves by
building contrary to her mandates.

It has been said that because man is a worshipping
animal he has a creator to worship and a soul to save.
But, dear reader, when you reflect that his desire to
lie to his friends, slander his neighbor, steal from the
unsuspecting, starve and torture his fellows, and

murder the innocent, is greater than his desire to worship his supposed creator, the former argument fades away into insignificance.

If one passion is divine then all must be. If one talent is Godgiven then all of them are. But as a matter of fact man has but a meager desire to worship anything but gold. Modern worship is mostly formal— little faith and less sincerity, and executed by a force of practice, usually for a purpose; therefore, there is no weight to the argument.

I will doubtless find myself homeless among the medical fraternity, because no one likes to have his business assailed; but falsehood must be sifted out of science regardless of individual desires or personal interests.

If I were writing in my own financial interest this would be a very different book. I would write lies to please, romance to flatter, and falsehoods for effect; for that would catch the masses ; for rugged scientific truth, stern and piercing, can be rel- ished only by the wise. Facts are alone for the freighted brain, and these are vastly in the minority; in these I trust for the success of this book.

I will now introduce the reader to " Stellar Dust," with the modest request that he, or she, read understandingly, without prejudice: weigh the arguments with care and judge according to the evidence adduced; then he or she will have done all an honest author could ask of an intelligent reader.

INTRODUCTION.

In a former volume, entitled "Evolution and Reproduction", which I published in 1889, was explained many points of interest in connection with this new philosophy, but after seven years of careful study I found it very imperfect. I therefore concluded to give to the public a more complete rendering of the subject, and to it add some of the most important points laid down in the former volume, since I cannot now tell when I shall have another edition published. And though I have been an enthusiastic student of Astrology for nearly a quarter of a century, yet I find no limit to the field of investigation, open to the untrammeled mind that can unfettered roam through nature's fields, blossoming with everlasting truths. I have an abiding confidence in the intuition of the human mind, notwithstanding its finite powers to unfold many of nature's most subtle secrets, which have hitherto been unrevealed to man, not only in connection with this world, but with the universe of systems. When philosophers, scientists and learned men extend their researches out into nature's open, broad and free fields, which is becoming a scientific mind, instead of ignoring the great universal natural laws, because they are called astrology and because a certain class of people have arrayed their forces against it to frown it down, then they may expect to

accomplish results in proportion to the efforts put
forth. But so long as they study effects beneath their
feet in searching for the laws which produce them,
just so long will they be wandering in the unproduc-
tive dessert of thought, all barren of results and
finally fall by the wayside, unrewarded for their toil.
In this, however, they only share the fate of many of
their predecessors, who surrender the scepter of life at
the very throne of success, had they only looked in
the right direction to behold the gems of truth ready
to flash out before their wandering eyes. What a
surprise would have been theirs, had they only turned
their eyes toward heaven, as they had so often been
commanded to do, and studied nature as a whole;
they would have learned much pertaining to her
subtle forces and become familiar with the secrets of
her mighty works, then how much they would have
learned to their own interest and to the advantage
of the human family, and how many facts in nature
they might have discovered long ago, and what
beautiful truths they might have scattered broadcast
to an anxious waiting world, and how much falsehood and superstition they might have throttled in
their embryonic form and thus made their lives both
profitable and glorious. But instead, they have
bolstered up stale dogmas, divine falsehoods, old saws
and ancient superstitions, which came to, and have
continued to curse the human family since they first
entered the mind of vicious man. It is painful to
contemplate the social and political conditions of the

world when compared with what it would have been had truth prevailed and falsehood been dethroned. Had the laws of nature been properly understood, generally taught and zealously observed at the proper time to quicken the human germ, that it might receive the full benefits of the benific, heavenly forces, which would have finally developed it into the full frutitions of manhood and womanhood. How different would be the condition of society today. Instead of vast sums of money being expended for the care of viciously insane, the hungry and the destitute, it might be spent to educate thousands of neglected youths and train their minds for lives of usefulness and personal satisfaction. But with all the power of human intelligence man has arrogated to himself. He is a frail, incapable creature, subjecting himself too much to the whims of the less capable, who have no aspirations above a petty personal motive. It is not the laws of nature that concerns their microcosm, nor the force of heaven that reaches their understanding. What do they care whence comes the gases which produce the wheat, sheep and cattle, so long as their tables receive their tri-daily supply of bread, mutton chop and roast ribs of beef, to sustain their wearing muscular tissues, to invigorate their mental organs and to refill their rapidly exhausting brain cells that they may have health and strength to jostle each other in the wild race for lucre.

What matters it to them whether God made the world in six days or that it was thrown from the sun

and found its orbit in six years, or by a slower process
it was formed by the accumalation of gases in six
millions of years, so long as her ample products are
forth coming at their command.

What does it matter to them whether our boys
are to be confined in asylums, locked behind bars,
or hold honorable positions in business, science,
literature, art, or profession, so long as their appetites
and passions are appealed, they love God and pay
the preacher.

Forethought is not a marked characteristic in the
human family, as proven by the lack of interest taken
in the subject of reproduction.

This is not a question of information, only
simply to gratify a longing for unknown facts of
nature, as the discovery of a comet, a distant star, or
the satellite of another planet, but it is one of
importance to all mankind. It is for the social good
of every nation and essential to the mental progress
of the whole world, individually and collectively; for
this reason it should enlist the attention and gain the
support of all, and especially those who are able to
think for themselves, but does it.

However, there has been a passing interest taken
in the origin of man since evolution has been placed
upon the spit. It has been roasted, cooled and
toasted again, assailed by its enemies and defended by
its friends, until its strong points will scarcely hold
its weak ones together, even in the hands of its most
powerful supporters, while its theological enemies who

have long and faithfully been drilled in the drama of creation as found in the first book of Genesis and purported to have been a divine revelation to man of the origin of life, condemn the Darwinian theory of evolution with all the vehemence at their command, while a very large percentage of the human family do not accept either as a correct solution, and are patiently waiting the promulgation of a more logical rendering of the problem, and one which is more in keeping with the present order of things than are either of the former doctrins.

And since there is a marked divergence of opinions among the honored, learned and great in regard to the beginning of life, the author feels justified in formulating a new theory which is supported by more facts than either, if not all of the known theories combined.

But doubtless many will deem it a presumption on his part to oppose the life efforts of the great Darwin, while others will declare it sacrilegious to doubt the statements of Moses, and perhaps sigh for the return of the Holy Inquisition to check the flood of free thought and thus preserve the institutions of fables, falsehoods and superstitions. But the dykes are down and the wild waves are upon them, washing the sands from beneath their uncertain foundations. It is only a matter of time when the great wrongs of the past and those practiced in the present will be righted, and might will not always be right, neither will superstition reign.

For years the author has been pleased to know that the inquisition of the "Holy of Holies" has been abolished, even in wicked Spain, and has had cause to rejoice that it is not now the rule of law on the free soil of America, where free speech is not altogether suppressed and where thoughts can freely flow without fear of priestly frown or the rack, dungeon or ax.

Though the vicious blade has ceased to vibrate in the jaws of the murderous gulletine to silence the voice of the thinker and reformer, yet tongues still continue to wag in defense of classified superstition. But their echoes are growing fainter and farther, and will finally be drowned in rejoicings of the people made happy that truth has come to stay.

BARBARIC LEGISLATION.

It can scarcely be credited that a man could have been found in the great state of California capable of framing a bill to suppress the freedom of speech, and especially the practice of a science in the free and easy going west. Yet there was just such a man, and furthermore there was elected a Senate of chosen men to represent and defend the liberties of the people of the great commonwealth of California, that voted upon that bill, passed it, and had it not been for the superior intelligence of the lower house, it would have become a law and thus put a check to free thought, scientific growth and disgraced the sun, set shore of freedom's sacred land.

For what? That superstition and falsehood might live and flourish off the credulity of an injured people.

HERE IS THE BILL.

"Any person who for valuable consideration or promise of reward undertakes to predict to another the future or reveal the past by means of cards, communications from the dead, the examination of any part of a person, of the dead or living, looking at the stars or heavens or representations thereof, planets or other bodies, heavenly or otherwise, or by any means not natural, or who prints or causes to be printed, or exhibits any sign or symbols intended to induce others to have their fortunes told, the past revealed, or the future predicted is guilty of a misdemeanor."

Now the important question arises: Who was it that introduced the bill? Was it an infidel, a spiritualist, a clergyman, phrenologist or a crank? Surely it was not an astrologer, but whoever it was, tried to disgrace the name of freedom and received more assistance in his heinous effort to suppress freedom of action than the wildest fanatic would have dared to guess. , But the march of science cannot be stayed by law, superstition, or neglect. It may encamp for the night to recuperate its ever increasing forces, but with the morning sun its unbroken lines will resume their impatient march.

But the foregoing is proof that the cloven foot still exists to protude from beneath the sacerdotal

robes of inspired divines, to crush human hope and
destroy the happiness of all mankind.   But when the
struggling masses once learned that the heavens rule
the development of the human brain they will then
have their visions clear to the fact that the black
robes of sacerdotal forms only cover the common
anatomy ruled by ordinary human brain, and the
majority of them doubtless are developed by the most
ordinary combination of planitary gases   From their
acts one would be compelled to think that they were
begotten in iniquity, reared in selfishness and
superstition, educated to deceive, mislead the young,
frighten the timid, plunder the poor and rob the
widow of her mite.

ZODIACAL SIGNS—HOW DISCOVERED.

The division of the heavens, as they were sur-
veyed and laid out by the ancient astrologers and
astronomers, for they are both one in practice in the
early history of the science. They were called signs
for the reason that they could not understand how
the stars so far away could effect the people of the
earth. In fact they thought the divisions were only
signs of what the physical development would be from
hereditary causes. They did not then know how the
Zodiac executed the great fixed laws of nature which
produced all things mundane, which they have since
been discovered to do.

In primitive days when there were no mechanical
devices, by which to note the passing hours of fleeting
time, it became necessary to devise some means by
which the important events of life might be recorded.

The handy clock and the more convenient watch,
and even the hour-glass were then unknown, and
though the semi-civilized people of those days were
very ignorant of all arts, science and mechanical
devices, nevertheless they discovered a method by
which to keep the records of the daily and nocturnal
events of human life. Their custom was to observe
the heavenly bodies and note their positions at the
time events occur. The position of the sun by day,
and the places of the moon and stars by night, and

in some crude manner make a record of them. They all knew when the sun rose, when it reached the mid-heavens and when it set. Then by practice they learned how to divide the quarters of the semi-circle into equal parts, and then tell the exact time of day by the position of the sun; then by selecting a bright star they could tell the time of night by its position, for the sun and stars ascended to the mid-heaven and set at the same rate of speed; therefore the rule that would apply to the sun would also apply to the planets and stars.

The sun and the planets were doubtless first employed for that purpose. Later on the fixed stars were brought into use. Venus and Jupiter being very bright and beautiful objects among the fixed stars to the rude children of nature, early became familiar to them; owing to their mutability they were curiously watched by the simple minded people of those far-remote unscientific days, before their influences were discovered. They observed the moon as she moved from star to star; they saw the planets change their places; they admired their beauty and loved their mystery. and what charmed them most they crown with the highest title of admiration of goodness and virtue. They felt the influence of the planets and knew they brought them good. They saw them sparkle and gleam in the depths of the blue firmament and called them shining angels. Further than this they knew nothing of those mysterious bodies, floating far away in their trackless rounds of endless space.

Those people are even now dubbed "star-gazers and sun-worshipers," and wise men sometimes laugh at the title thus bestowed.

Yet is there anything higher in the form of worship, holier in the desire, more beautiful in its simplicity, or more practical in its results than adoring a living truth? Let those who worship an ancient myth, a flaky wafer, a glass of wine, the spirit form, and laugh at astrology, reflect.

Astrology doubtless was the highest form of worship ever known to man. From Jupiter came the word Jove or Jehova; from Saturn was derived Satan. The Lord of hosts (of stars), The Most High fixed stars, shining angels, ministering angels, angels of glory and angels of light, angels of darkness, seven angels, and swift winged angels, are all astrological terms as applied to the fixed stars and planets; but they have all been misinterpreted for a purpose, but eventually their true meanings will be understood.

A writer commenting on the number seven found so often in the Bible refers to it as a sacred number, because, I presume, he did not understand why it was so profusely employed in that connection; but had he looked far enough to have discovered its origin he would have learned that the number originated with the seven planets : Saturn, Jupiter, Mars, Venus, Mercury, Sun and Moon, all which were then known to the ancient astrologer.

The following is a partial list of the things numbered by sevens: The days of creation were seven; the years of famine and plenty were numbered by seven; there were seven days of the week, and every seventh day was the sabbath of rest; after seven times seven years came the jubilee; the feast of the unleavened bread and the tabernacles was observed seven days; the golden candlesticks had seven branches; seven priests with seven trumpets surrounded Jericho seven times, and seven times the seventh day; Jacob obtained his wives by a servitude of seven years; Samson kept his nuptials for seven days, and on the seventh day he put a riddle to his wife. He was bound by seven green withes, and seven locks of hair were shaven off; Nebebuchadnazzer was seven years a beast; Shadrach and his two companions in misfortune were cast into a furnace heated seven times more than it was wont.

In the new testament everything occurs by seven; in the revelations we read of seven churches, seven candlesticks, seven spirits, seven trumpets, seven plagues, seven vials, seven seals, seven stars and seven headed monsters, and is pretty good evidence that the ancients, whom people now suspect were Christ worshippers, were star worshippers instead, with the Christian doctrine as a side issue. By the practice of observing the heavens for the purpose of recording events, they became familiar with the positions of the fixed stars and the motions of the planets.

Occasionally all of the planets would be invisible by being beneath the earth during the night time, therefore could not be observed for the purpose above mentioned. It then became necessary for them to

single out some of the most prominent fixed stars to
be observed in the absence of familiar planets, after
which it was discovered that a new influence existed;
then by careful observation it was noticed that the
general influences supposed to emanate from fixed
stars extended over 30 degrees of space, and that all
children born at the rising of any one of the twelve
divisions of the heavens partook of like developments
and general characteristics. Thus they discovered a
very convenient and reliable system for finding the
time all important events occured, and also the zodiacal
influences in the brain. This observation continued
till they had discovered twelve distinct forms and
characters produced by the twelve zodiacal divisions
which they called houses, or mansions of the heavens.
Hence the quotation: "In my father's house are many
mansions."

As language was very imperfectly understood at
that time, and embraced only a simple dialect, then
the weak understanding of the ordinary people
made it necessary for the teachers of those days to
resort to illustrations, and draw comparisons from
what was known of their mundane surroundings. In
order to impart the heavenly knowledge the more
thoughtful ones had obtained by observation, the animals of the forest and the beasts of the fields were chosen
to symbolize the respective houses of the heavens.

ARIES.--The ancients noticed that all persons
born at the rising of the first divisions of the heavens
were irritable, quick tempered, active and combative.

To illustrate this character they selected the animal which they thought would answer the purpose best. As the ram was known to possess the foregoing peculiarities he was chosen to symbolize the division of the heavens which produced the effect in man and called it Aries, the Greek word for ram.

TAURUS.—The next division of the heavens produced a short, stout body, a mild, pleasant disposition until aroused, when he became as furious and terrible as the bull. He was therefore chosen to symbolize the second division of the heavens and called Taurus, the Greek word for that animal.

GEMINI.—The division following produced a quick, active, pleasant disposition; a person fond of climbing to high places. It being a double bodied sign, the ancient astrologer symbolized it by the twin kids, but later on they were substituted for the human twins, because of the fine feelings and sensitive nature it produced. It is called a double bodied sign since the moon in that sign at the birth of a male child causes a plural marriage. The sun in that sign at the birth of a female child causes a plural marriage.

CANCER.—The next division in the circle of the heavens is Cancer, so named because it produces a mild, inactive, non-aggressive character, who would shun an enemy, but fond of the water, and like the crab, prone to retreat; hence the crab was chosen as a representitive of the character produced by that division and elevated to the heavens.

Leo.—As the lion was symbolical of the disposition produced by the succeeding division he was immortalized in the heavens, and called Leo.

Virgo.—The next division produced a modest character, pleasing in manner and possessing an unusual degree of refinement. For this reason the fairest and purest of their tribes was chosen to symbolize the division, and called Virgo, meaning the virgin.

Libra.—The following division produced a mild, pleasant, inoffensive character, with a gentle and yielding disposition, unvarying in manner and courtesies. For these reasons this constellation was symbolized by the balances.

Scorpio.—All persons born at the rising of the succeeding division were pleasant so long as they were not molested; could have their own way in all things; but no sooner were they crossed or offended than they became bitter and sarcastic; their tongues would sting like the sting of a scorpion. For this reason the scorpion was chosen to represent the disposition of the person ruled by that sign.

Sagittarius.—The pecularity of the person born at the rising of the next sign could not be illustrated by any living animal. Therefore they were compelled to invent a figure for that purpose, and since all persons born at the rising of this sign were fond of horses and hunting they had to combine the forms of two animals, the man and horse. Consequently the Centaur was invented and called Sagittarius, meaning half man and half horse, and thus the division of the heavens which produced the effect was symbolized.

CAPRICORNUS.—The succeeding sign of the zodiacal belt produced a quiet person, having thin beard and a peculiar springing gait, giving at the knees when walking, and shaking his head when talking; changeable and somewhat fickle. Some of these peculiarities were seen in the goat; he was therefore selected to represent the person born at the rising of that zoadical division and named Capricornus, meaning the watergoat.

AQUARIUS.—The succeeding division in the zodiacal belt produced a degree of refinement, fondness for flowers, pictures, paintings and decorations; as the flowers were the only means they had of gratifying that taste, they cultivated, watered and watched them; thus the division was symbolized by the waterman.

PISCES.—Pisces the last sign in the circle of beasts, is represented by the fishes, because that division of the belt when rising produces a fondness for water, swimming, boating, fishing as well as drinking. This completes the Zodiac, which means a circle of beasts.

CHAPTER I.

MATTER.

ATOMS.

'Tis glorious to gaze on the firmament,
And study the forces that be,
To watch from afar each planet and star, ·
That creates all life we can see.

But whence come those atoms of matter,
Which are sent from the planets to earth?
To give force to the seed of the flower,
And to animal life give birth.

Man has struggled this problem to master
For thousands of years, they say,
But the source whence sprang those atoms
Will remain hidden forever, and aye.

Though man can analyze physical matter, and separate it into its component elements, and also name the gases of which it is composed, still matter is a mystery to man. Whence it comes he does not know, neither can he find out. Its original condition before it resolved itself into suns, worlds and satellites is supposed to have been gaseous, but how they were created is not known. Some scientists think that gases have always existed; perhaps it is true.

I have no serious objection to endorsing that view of the matter myself, since it is a very safe position to accept, because of no immediate danger of being dislodged therefrom. Gases may have always existed, but it matters not whether they have existed always or only half that length of time. If the latter be true, I cannot tell just when the first cloud began to form.

How each atom of matter was endowed with its special functions cannot be known, and why they will unite with each other will continue to be unexplained. The physical process of growth may be known, but the chemical action never ! That is a point too subtle, and a cause too remote for the grasp of the human mind; nevertheless, some will speculate on causes as long as they are able to think, reason or plan, or till nature renders her most profound secret to man. Most people are satisfied to let the subject rest, content with believing that all nature is the work of an unknown and unseen hand. The thinking mind is always agnostic and wants to know why. He is willing that God should have all the honor there is in the office, when he is satisfied that it belongs to an intelligent existing designer and creator. But until that point is settled he will ponder in doubt and continue to search for causes until the end. So far as man can penetrate the secret there is no power superior to the chemical action of matter, which alone is able to create force, motion, worlds and life.

When the human mind is able to grasp the laws which cause two or more atoms to unite in creating forms, animal or otherwise, he has comprehended all there is of God, solved the problem of life, the growth of bodies and the creation of worlds. To comprehend the smallest act of nature is to understand her most profound laws and stupendous works. Man knows something of the planet on which he has built a temporary home and has an imperfect knowledge of the solar system; but he cannot conceive the magnitude of the mind required to plan or the strength of the arm necessary to put into motion even the bodies which revolve in such perfect order about the genial sun.

Probably man can comprehend a being able to watch over this little world of ours, count the hairs of the head, number the sands of the sea-shore with little help, but he can know nothing of the being, able to create boundless space and people it with flaming balls of firey matter, dead bodies and opaque worlds; not to mention the creation of gases.

A glance into starry space is enough to engulf the thoughtful mind in a sea of reflection, and to reveal to man his utter insignificance. Then, instead of extolling himself to the tenth heaven, there to sit beside the architect and builder of the mighty universe, he will feel more like clothing himself in sackcloth and ashes, and seeking the society of one of heaven's discarded angels. if he has enough vanity left to think himself worthy the society of a member of the royal family.

CHAPTER II.

DISTANCE.

One of the very difficult obstacles to surmount in the path leading the human mind up to the point of accepting the sublime philosophy of planetary influences as set forth in astrology is the question of distance.

It is generally supposed that the planets and fixed stars are too remote from the earth to have any influence in the human organism, even if they do throw off gases, notwithstanding the fact all thinking minds know the contrary to be true, nevertheless they dispute astrology.

It has been calculated this earth is about 92 millions of miles from the sun, a distance no one can comprehend, but when we read of a death from a sun-stroke no one doubts the fact, but the thoughtful man ponders long over the same, trying to solve the problem for himself, but he is sorely puzzled to explain how it was done, to his own satisfaction, since he cannot understand how heat can be transmitted so great a distance with such a violent force simply by radiation, even if the sun rays did not become cooled in making the transit, how the earth could receive them with so much violence after their force is nearly if not quite spent, as one might well suppose, as they would be after traveling that distance, is a thing not easy to understand.

To pass through that space at the rate of 250,000 miles per day would require about 365 days to complete this immense journey, even if there was no loss of energy or heat during its transit. The earth would not receive a sufficient number of electric volts to produce the effect reported, therefore it is fair to conclude that the theory of radiation is wrong; but it might be reasonably accounted for by the force of attraction.

By attraction of gases may be brought to the earth with a much greater force than by radiation, because the nearer the rays of heat approach the faster they would advance.

But even if there was no difference in the force produced on the earth by a single atom of matter by these two laws, attraction would have the advantage in the number of atoms collected at a given point on the earth, since attraction converges, and consequently would draw them together, while radiation diverges, therefore would scatter, light heat and gases; so that by the time they reach the earth they would have but little force left to produce any effect.

The more atoms attracted to a given point the greater would be the effect produced by them. Therefore attraction is more reasonable than radiation, regardless of the phenomena produced. Then if the earth could attract sun forces she could also attract planetary gases. Another evidence in favor of attraction is that the planet Mercury supplies the gases which produces the organs of causality. When Mercury is at the point nearest the earth or about 50,000,000 of miles away he is in his weakest position except the opposition, and produces the least visible effect in the brain.

When he is nearly twice that distance from the earth he is then in his strongest position and produces his greatest development of the brain. The explanation for the different effects is as follows: The earth attracts gases from the sun, consequently there is an immense current passing between the sun and the earth. When Mercury is passing through that current his gases are so dissipated by the sun's rays that a child born at that time cannot receive it, consequently causality does not develop. But when Mercury is at his greatest elongation he has passed out of the sun's rays, and therefore his gases are attracted to the earth in his purest state, so the brain of the child born at that fortunate time can and does receive and utilizes them in producing brain matter.

The foregoing facts I have many times proven by locating the position of the planet Mercury from the development of these organs, which is done simply by the touch. When causality is large it is safe to say that Mercury is free from the sun's beams. If the sun radiates his forces to the earth then it would necessarily follow that he would radiate them the same distance in all directions from his center, consequently Mercury would never be free from his beams, and the farther he would be from the earth the greater would be the depths of solar rays his gases would have to penetrate to reach the earth, and the more they would be absorbed by the sun's forces in their journey, furthermore the farther Mercury is from the earth the greater would be the divergence of his rays, consequently a less number of them would reach the earth, therefore the farther Mercury is from the earth the weaker should be his effect in the brain, which is not true until he reaches an afflicting point beyond the sun. When the sun's rays again impede the forces of the Mercurial gases, when his force is again weakened. Thus it can be seen that it is not distance which produces the different effects in the development of the Mercurial brain, but the attraction of the earth.

Mars, Venus, Mercury, when at their nearest approach to the earth are not nearly so far away as the sun is all the time, therefore if the earth can attract the sun's forces all the time, she should be able to attract planetary gases at least, part of the time, even if distance does interfere.

If the sun can radiate his heat and chemical forces 92,000,000 of miles and produce the above named effect the planets could not, for their power of radiation must be extremely limited, owing to their temperature, yet they are known to affect the earth and all life on its surface. If the gases were forced here it is not an easy matter to understand why the atmospheric pressure would decrease with the square of the distance from the earth. If any difference the forces would work the other way; for radiation of heat is one atom pushing against another. Oxygen is required to produce fire, when wood is being consumed, oxygen rushes in the to the assistance of the carbon, thus crowding all of the liberated gases outward. The greater the fire the more force would be produced by the oxygen rushing in, and

the faster would the gases be crowded into space; but
as fast as they would get room they would decrease in
motion, even if they did not stop altogether. Then,
after they had entered ethereal fields they would not
be likely to plunge on into a more dense atmosphere,
like that of the earth's, with a force so violent as to
destroy life in man and beast and vernal vegetation.
Scientific men have arranged it among themselves to
have the atmosphere retained at the earth's surface by
attraction, while light, heat and gases are affected
differently; but I protest against such absurdity, and
place them all under the same law, the law of attraction. The waters of the ocean are greatly agitated all
the time. The cause of its unrest is attributed principally to the attraction of the moon. The moon is a
much smaller body than the earth, but her admitted
attraction is sufficient to lift the water in midocean
one foot heavenward. If the moon, though only one
fiftieth size of the earth, can affect ¡the ocean to that
extent by attraction, then it is not unreasonable to
infer that the earth can attract gases from the heavenly
bodies. The moon is said to be a dead body; that is,
she has no moisture; therefore, life cannot be created
on her surface, nor could it long exist if it were created.
The want of moisture is owing to the fact that she has
no affinity for both oxygen and hydrogen gases, the
union of which is necessary to produce water; but
doubtless she has a strong attraction for one of them,
the effect of which is to disturb the waters of the earth
by trying to separate those gases.

Venus is a much larger body than the moon, and, logically speaking, she ought to exert some force on things mundane, even though it is not so sensibly felt as the forces of the sun and moon. Her attraction for some of the earth's gases might be just as powerful as that of the moon, but not for oxygen or hydrogen, and therefore her effect is not so discernible on the face of the waters.

Owing to their immense distance from the earth it is quite impossible to judge the effect one planet has on another, by any variation from their usual course. At most the effect can only be very slight. The influences of the sun and moon are plainly discernable on the earth; but not so with the planets. Their effects are only noticeable in the development of the brain, which is the only way anything definite can be learned concerning the influence on the earth. The remote positions of Uranus, Saturn and Jupiter from the earth, their temperature, etc., positively forbids the power to radiate their gases the immense distance through which they must pass to reach the earth, and yet their gases are known to be here. Therefore it must be conceded that they reach the earth by the force of attraction.

The distance of Mars from the earth varies greatly. His nearest approach to it being about 50 millions of miles, his most remote point being about five times that distance, or about 250 millions of miles. Yet astrologers have discovered no difference in his influence when at different points in his orbit at the time of birth, which would not be the case if he radiated his forces. He is just as potent when 250 millions of miles away as when only 50 millions of miles from the earth. In making his transit Mars is said to be more malific when passing the conjunction of the sun, than when passing in opposition to that luminary. Of this, however, I am not sure, but if it is true I have no explanation to offer for the paradox, but will leave it with those who have made the observation to clear up the mystery. Mars is five times the distance from the earth when making his conjunction transit than when he is passing the opposition, and why he should be stronger when he is farther from the earth, I cannot explain by any hypothesis.

Mercury, like the moon, is also said to be a lifeless body. Some scientific men say that he is hot; yes, red hot, being so near the sun, only 37 millions of miles away; others say he is cold as ice and hard as lead.

"Being so old and a long time dead
He is petrified, I think, they said."

I sometimes think that scientific men and religionists enjoy studying subjects they know nothing about, and upon which they never can become informed. Perhaps it is because their views on the subjects, however wild and illogical, cannot be successfully disputed. They therefore cannot be vanquished.

When men will study the planets in space without first learning all they can about them at home, savors of the unmentionable, and when they will dispute known facts about the planets for selfish reasons, there is room to doubt their honesty or intelligence; yet these men have great influence in society.

The author once read the horoscope of a young lady just before she visited Lick's observatory, on Mount Hamilton, California. After taking a view of the mighty planet Jupiter through the immense telescope, she asked the man at the glass if there was any truth in the planets having influence in the human brain. "No," he replied, "Astrology is an ancient myth." And the young lady, right in the face of personal evidence, believed the professor because he was drawing a big salary.

Very little is known of the temperature of the planets, and very little else except their motion and influence in the human brain. If the fires of Mercury have smouldered and gone out, and he has no internal heat, and must depend upon the heat of the sun to operate the quicksilver column, his temperature does not run very high, and unless his power of attraction for the sun's heat is proportionately much greater than that of the earth, he is not very hot, and the radiation of the sun's rays could not affect his people disastrously; but so far as his physical conditions affect the mental condition of man on this earth, it matters not whether his people bathe in lakes of molten lead, or live in huts of ice; whether his gases are hot, cold, moist or dry. Their effect on the earth would be the same under the law of attraction, since the earth would only attract her affinities from that body, and their temperature would be the same on reaching the earth, whatever their condition might have been before leaving that body.

There is no evidence against the magnetic powers of the earth being able to attract gases from any and all celestial bodies regardless of distance. Venus is very potent for good and, comparatively speaking, she is very near the earth. Uranus is very potent for evil, and on the contrary he is very remote from the earth. But if by some sudden and mysterious force these two planets should exchange places, their chemical influence on the earth would not be changed in the least degree; for the attraction of the earth is sufficient to reach her affinities at any distance.

Saturn, Jupiter and Uranus are opaque bodies, and could not radiate their gases for the want of heat, even if the sun could throw his heat 92 millions of miles; but the fact of the matter is, none of them do, notwithstanding the fact that their gases are here and the source whence they came is well known to all students of astrology. Astrology was practiced thousands of years prior to the discovery of Uranus, but so closely did the ancient astrologers observe the planetary effect in mankind that they discovered the influence of Uranus ; but not knowing whence it came, they attributed it to the dragon's head and dragon's tail, and part of fortune, the two former names being given to the moon's nodes; the dragon's head being the moon's north node, and the tail the south node; and the part of fortune a given number of degrees of areas, three imaginary points in the heavens whence

originated no influence whatever; so when they found
a person whose horoscope they would not reveal their
characters without the employment of these imaginary
points, they were brought into use and were made to
account for the influences of the unknown planet.
Nor was the fallacy of this practice observed till after
the discovery of Uranus, which occurred in 1781; but
since that time astrologers have been gradually casting aside those old mythological notions till now
there is not an astrologer of repute who observes the
dragon's head, tail, and part of fortune; but all have
discarded them as vagaries. These facts are over-
whelming in favor of mundane attraction and planetary influences, working in the human brain. If the
above named forces had not existed, the ancient mas-
ters would not have employed imaginary points in the
heavens to frame a theoretical science, when there are
so many bodies in space which would have answered
their purpose just as well, and which they could have
employed with much more reason if they were only
framing a hypothetical system of divination. But no—
their system was based on scientific principles; but
imperfect, because all of the planets were not known
to them. And Uranus, being so powerful when in a
certain position in the heavens, they could readily
recognize his effect in many of their subjects; consequently they knew his influence existed, although the
planet was unknown.

Just here let me state a mathematical proposition:
If a spherical magnet, 7 inches in diameter, can

attract a fluid 1000 miles, how far can the earth attract gases, presuming both to be of equal power? Answer: 11,088,000,000, nearly six times the distance to Uranus. I give this because the brain which is about that size is believed by some to possess that power. It has been stated by some that the Oriental psychics can send and receive mental messages hundreds of miles, which I have never disputed, because I have known thought to be sent nearly 100 miles distant, for which reason I cannot dispute a greater distance. I was personally acquainted with a lady in Kansas, whose husband was in the army during the rebellion. He was stationed at Fort Scott, a distance of about 100 miles from home. This lady could always tell when her husband was coming home on a furlough, and so could her neighbors, for she told them beforehand. She could give no reason for her knowledge. She only knew it was true. It a brain can receive a message through 100 miles of space filled with atmosphere, why not further? and further what is to prevent the earth from attracting gases through ethereal space.

CHAPTER III.

WORLDS.

It will scarcely be expected of any one writing on scientific subjects to enter into a detailed account of nature's methods of creation, even after she has so far advanced her work as to create gases. Nevertheless, I shall offer a theory which may or may not be original, but so far as my knowledge extends it is not very ancient. I do not present it for the purpose of trying to establish a new theory only so far as it seems necessary to explain why celestial bodies harmonize with each other, and how they produced life on the earth. Without some plausible reason for astrology many people would reject it without investigating; but with the necessary explanations, accompanying practical demonstrations, it will be readily accepted by them.

I know of nothing just like it in all the theories introduced to the public, but be it original or second-handed, it matters not to the author. He will present it to the reader of Stellar Dust as the most plausible, not to say scientific, in his judgment, of any theory yet brought forward on the subject of world-building. By some scientific men it is believed that confusion reigned throughout the universe prior to the creation of physical matter, and that out of the chaotic condition of gases the earth and other bodies evolved; but doubtless such was never the condition of matter. Nature is a stranger to confusion. No matter how complicated her

laws may be they never were confused. Should confusion once reign it could never be dethroned; for confusion implies the absence of law and order, without
which no physical results could be produced. What in
some cases might be considered chaos, is perfect order,
and what is said to be a violation of a natural law is
only a confirmation of that law; for natural laws are
not to be violated. But the ruling of one law might be
very different from that of another, and the changing
of their forces would produce different results in their
work. So far as man can judge by observing the atmosphere it is a confused body of gases, but from the
result of nature's works in the production of physical
form he is able to judge a true condition of her elements.
Nature's works are always perfect as the laws which
create them are perfect, and all laws are true to the
nucleus which they create, be it a flower, tree or world;
therefore the conclusion must be that nature and order
are synonymous terms.

Every atom that flies in space is alive with force
and endowed with more or less sensibility, if not intelligence. There is sound reason for thinking that
the different gases, 75 or more in number, are as perfectly organized in their chemical relations to each
other, while in their original state, as they are when in
physical forms. Though each atom of matter, while
in a gaseous state, is free to move about in its own
system undisturbed, nevertheless it cannot lose its
individuality nor its relationship to the nucleus to
which it belongs. Why one atom of matter will attract

another and repel the third is a deep mystery, but they do, and in so doing they form associations, and thus produce life of every kind.

A narrow survey of our immediate surroundings confirms the truth of the statement that every form of life is formed by the specific combination of gases. All the fruit trees, shrubs, plants and flowers, differ from each other in appearance, texture of wood, form of leaf and odor emitted; but just what suggested the original nucleus still remains a mystery.

Motion is an innate function of matter, consequently gases must have been in motion before they gathered into separate bodies. In moving they form circles about different centres. These currents may have caused neuclei to form while rapidly moving in their unvarying courses. The relation of the fixed bodies of gases was the chemical relation they bore to each other, for each gaseous body exercised an attractive and repelling force. Each body of gas contained a nucleus different from all other bodies, and about which they gathered their affinities. The number of combinations which it is possible to be produced, even by the gases known to the people of earth, is innumerable. And the unknown number might increase the combination many fold. Thus under the law of attraction each and every atom of matter in space was enabled to find its place in the mighty universe of systems. When a nucleus was established its individuality must continue through all time. If it could hold its position n space and sustain the body of gases collected against the attraction of all other nuclei for millions of years, the transformation from a gaseous body to a physical form would not change its individuality nor power of attraction. Thus, once organized into systems, the universe of worlds must continue forever in their established relations to each other without clash or disturbance.

In this way would every celestial body become a central force, and thus would the smallest be able with the largest body to hold its place in the vast economy of nature. Thus was matter drawn together and organized into systems of gases. In the course of time a change came over the conditions of the nucleus of each gaseous body which wrought in them a greater power of attraction than they had previously possessed, but what produced the change I am unable to explain, except it was motion. The greater the motion the denser became the body. As the force of attraction increased, physical matter was produced; as physical matter formed, its density increased and the play of friction began. The stronger the attraction the greater the friction till gradually it became so great that heat was generated, which condition continued till the gases in the heavenly bodies were transformed into physical matter and reduced to a molten condition. The heat continued till all the mass became as dense as the force of attraction could make it, when the play of friction ceased altogether, after which the cooling process began. The heat radiating for a long period of time, finally the surface of the earth became cool, and later on a crust formed, when the once luminous became an opaque body. As the crust became thicker, the surface became cooler until it reached a proper degree of temperature to produce incubation, when life was generated. As the crust became too large to snugly fit to the molten matter within, it was forced up on certain lines which form mountain ranges extending from north to south. The motion of the earth upon its axis doubtless caused the crust to be thrown out upon those lines instead of breaking on the equatorial parallels, and thus formed mountain ranges running east and west.

After the earth had become cool enough, the heavy clouds which surround it began to condense and fall in rain, until vast bodies of water covered the entire surface of the earth, for it was spherical. The weight of water had a tendency to depress it, until finally the crust gave way on certain lines and crowded the surface out on parallel lines. The centrifugal motion of the earth, in connection with the weight of water on the surface, caused the mountain ranges to form as they are now seen. Internal explosions, as some theorists explain, would throw the crust up into one huge pile, like the peaks which are seen on all mountain ranges, instead of in chains.

Be the foregoing true or false, it is the only theory which will reasonably account for the different chemical constituents of the planets and fixed stars as they are laid down in astrology. For this reason alone, I shall defend the theory until a better one is brought forward.

CHAPTER IV.

PLANETS.

Sun.—Beginning with the sun, center of the solar system, which is said to contain 500 times as much as all the rest of the matter within the limits of this mighty system. It has been computed to be about 768,000 miles in circumference. Traveling at a speed of thirty miles per hour, it would require only about forty days to encircle the earth; but to girdle the equator of the sun would require nine years, Sundays not excepted, traveling at the same rate of speed. Apparently he is stationary, but revolves on his axis in a little less than 26 days, at the astonishing velocity of about 88,000 miles per hour. He is supposed to be the giver and sustainer of all life, which is only true in part. His density is computed to be 25 hundredths, as compared with the earth, or about one-half as heavy as water.

Mercury.--Leaving the sun, and passing out 37 millions of miles into space, the orbit of Mercury is reached. Here we find a planet 19 times smaller than the earth. He is moonless, without atmosphere or life. His diameter is 2962 miles. His density, compared with that of the earth, is 1.24. He requires 88 days to pass through all the signs of the zodiac in his journey about the sun. His velocity is 110,000 miles per hour. He revolves on his axis in 23 hours. The gases emanating from this planet produce intelligence by developing the phrenological organs of casualty. The color of this planet is red, with a bluish tint.

VENUS.—Leaving the planet Mercury, and passing out toward the zodiac, the orbit of Venus is reached at a point 68 millions of miles from the sun, where is found the "goddess of love." She is computed to be about the size of the earth. Her nights are forever dark; perhaps that is the reason she is not inhabited, for lovers will not woo where the moon does not shine. If she possesses human life it is of a low intellectual order, far inferior to the people of earth, and that is not saying much for them. She has an atmosphere, though less dense than that of the earth. Her diameter is 7510 miles. Her density is .92, about the same as that of the earth. Her diurnal revolution is about 23 hours. Her motion on the equator is 1100 miles per hour. She moves in her orbit at the rate of 77,057 miles per hour. Her color on a bright, clear night is a mixture of silver and gold. She produces the gases which develop the love nature, fine arts and refinement in the people of earth, and was therefore called the "goddess of love" by the ancient "heathen" star-worshipers, but this is what the tramp poet says about star and other worshipers. The reader can take it for what it is worth:

WORSHIP.

"To gaze on the stars as they shine in the skies
And reflect their bright lights and their forces.
And bow to their beauty and influence, is grand
To those who have sense above horses.

"But to worship a myth somewhere in the skies
And mumble a senseless phrase
Over a cracker and glass of wine
Is weak, even for a horse that brays."

The next resting place in our outward march from the sun is the

EARTH—About which very little is known. Some scientific men say that she has an opening clear through her center from the north to the south pole; that she is inhabited and her people live on the inside of her shell. Others say that she is solid as a brick, and her people live on the surface. The latter I know to be a fact—in some cases, at least. She has one moon, and revolves upon her axis once in 24 hours, 48 minutes and 48 seconds. Her diameter is 7912 miles, and her distance from the sun is about 92 millions of miles. She has an atmosphere and produces life. Her density is one. Her orbital motion is 65,533 miles per hour.

MARS.—We find the orbit of Mars 144 millions of miles from the earth. He is next in size to Mercury, and six times smaller than the earth. He has two moons, a rarified atmosphere, and may produce life. He is agricultural in his habits, and mineral producing in his resources, which facts are established by the discovery of an irrigating canal which extends across his entire surface, and three sluice boxes in the mining districts. His diameter is 4920 miles. His density is 1.24, which makes its surface harder than that of the earth, but their plows are made of better steel and their horses are stronger. His diurnal revolution is completed in 24 hours and 37 minutes. His velocity at the equator is 628 miles per hour. His velocity about the sun is 53,000 miles per hour. His year is 684 days, and his color is fiery red.

JUPITER.—This is the largest planet in the solar system. He is 490 millions of miles from the sun. He has five moons, consequently is blessed by moonlight every night in the year. He is young in maturity, and is not supposed to have reached a point in his development that would enable him to produce a family, and is therefore supposed to be uninhabited. The diameter of this mighty speck of matter is said to be 88,390 miles. His density is estimated to be even less than that of the sun, which, if true, would involve a very difficult problem for astronomers to solve, since that would make him a luminous body. He is known to be opaque; but perhaps he is in his gaseous state, and has not yet reached his luminous condition, and therefore is liable to blaze forth at any time, and thus add another beautiful sun to our system. I think, however, he has passed the bright days of his youth, and has entered, if not passed his fruitful period, and since hs sustains such perfect relationship to the zodiac and the planets in our solar system, I must conclude that he is inhabited, not by a low order of animal life, but by a superior race of beings, which

I judge to be true from the number of moons he possesses. If each of these is as powerful in developing brain matter as the earth's moon is, his people have at least five times the amount of brain power that the people of earth possess. Again, if his attraction for zodiacal gases is in proportion to his size, his people must be 1200 times larger than the people of this earth. If a Jupiter man, in his red shirt, was to step on this planet, with trumpet in hand, the whole Salvation Army would have a regular Hallelujah meeting without notice, and that God had set up his kingdom upon earth; but since there would hardly be room enough in our streets for his big feet, it would not be wise to extend an invitation to the Governor of Jupiter. The diurnal revolution of Jupiter is completed in 9 hours and 55 minutes of earth time. His solar speed is 28,744 miles per hour. His equatorial speed is 27,985 miles per hour, which is nearly 28 times faster than the earth moves at her equator. His year is 12 mundane years. His seasons are three years each. His color is a soft red. He was worshiped by the ancients as the God of Justice. He is the greater benefic, and the redeeming feature in the plan of life, if there was any plan in it.

SATURN.—Saturn is the next orb encountered on our outward march toward the pleaids. He contains an immense bulk of matter, being only one-sixth less than that of Jupiter. He is one thousand times larger than the earth. He has two marvelous rings, and eight beautiful moons, sublime in his aspect, but exceedingly malific in his influence. He is an opaque body, and said to be light as cork. His density is said to be only .12, which, however, is not true, or he too would be a luminous body. He has an atmosphere, and I judge has life on his surface. His diameter is 71,904 miles; his diurnal motion is 10 hours and 29 minutes. His equatorial speed is 21,528 miles per hour, his year is 29½ mundane years. His seasons are nearly eight years long, and his orbital motion is 21,221 miles per hour. He is the greater malific.

URANUS.—In the immense flight across the fields of space from the sun, travelling at railroad speed of 720 miles a day, we reach the orbit of Uranus after a journey of more than seven thousand years, through 1,800 millions miles of space. Here we find a planet 100 times larger than our earth. His diameter is 33,000 miles. His density is .18, and therefore should be luminous. He travels in his orbit at the rate of 30,787 miles per hour. He has a pale ashy color and possesses five moons. He is sensitive, original, mechanical and intuitive in his influence in the human brain, and may or may not be inhabited.

NEPTUNE.—Speeding outward to the extremity of the solar system, through 27,747 millions of miles of space, after a continuous journey of 12,000 years, we reach the orbit of Neptune, where is found a planet perhaps 50 times larger than our earth, and possessing four moons. He too may have an atmosphere, and also possess life. His diurnal motion is unknown, but

his orbital motion is estimated to be 11,958 miles per
hour.   His diameter is?   His density is said to be .17,
which would make him luminous also.   Consequently
we ought to have three more suns in our solar system,
if the theory of creation set forth in the foregoing
pages is correct and their density known.

CHAPTER V.

CENTRIPETAL FORCE.

The philosophy of this doctrine was laid down in the following form: The sun was once a mass of molten matter, and also a mighty magnet, possessing rotary motion. It completed its revolution on its axis in a little less than 25 days. This motion created centrifugal force. Its attraction is called centripetal force. These two forces were continually in operation, the one to throw apart and the other to hold together, solar matter—the stronger to prevail. In the case in question centrifugal overcame the centripetal, and mass after mass was hurled into space, and eventually became planets.

According to the aforesaid theory, after a mass of matter was sent flying off from the sun, centripetal held it in check, and finally brought it to a circular motion about that body; but it not being strong enough to recover the detached fragments, it continued to follow the same path from that time down to the present. In substance, this is the theory advocated by some astronomers. It has long prevailed, patiently awaiting the arrival of a more reasonable one to supercede it. It is a good theory to believe outside the school of logic.

The Explanation.—The foregoing theory is uni-

versally taught, and owing to its extreme popularity
is not a pleasant subject to attack.   In fact it is quite
a serious matter to assume the responsibility of trying
to overthrow a theory so well founded in the minds of
all, and one so universally taught throughout the
entire world as this one is.   If universal recognition
of a supposed truth is conclusive in establishing it as
a scientific fact, it would be useless to question this
one; for it is recognized by all nations as the cause of
planetary motion.   To assail it will no doubt incur
the displeasure of all its friends, who will receive it as
an insult to science and an offence to the highest stan-
dard of intelligence, if not as an absolute crime. for
which the offender should be punished.   However, it
is laid down in legal lore that a man is innocent of an
alleged crime until proven guilty; though all the world
may be arrayed against him, the fact of his guilt must
be established before sentence can be pronounced on
him.   The knowledge of these facts is encouraging,
since there is nothing to be feared from the tribunal of
justice, whatever may be the opinion of the lobby, for
the reason that the theory is not in the least degree
logical in its conclusions, nor is it satisfactory to any
thinking mind that has not been thus instructed in
its youth and learned to believe it as it has learned to
accept many other teachings because of their current
worth.   The forces in question, however, are not established by scientfic facts, sound logic, nor even
by a process of mild reasoning, but simply by common
consent, which I must admit is of itself not a very

domestic force for a lone combatant to oppose; but knowing the weakness of the enemy's guns, and the vantage ground of their foe, I hesitate not to enter the conflict for championship of theories. The vindicator of these forces, in order to make any kind of a defense, must first show that the solar system is an individual entity of stellar matter, possessing an independent force, deriving all of its powers, just or unjust, from a solar orb. He must also show that it is a self-supporting factor of the mighty universe, and that the sun is the parental center, which once possessed all the matter now belonging to the solar system, which will not be an easy task for him to perform. Until the foregoing points shall be established he will have no permanent basis on which to rest his defense, and after it is accomplished he will be wholly at sea without boat, paddle or compass. In reality I think the theory has no staunch defenders among the thinking class of people, if, indeed, it has any. It is taught to fill an otherwise blank page in natural philosophy. This theory does not appeal to human reason since there are no facts to support it. It at once becomes legless, and therefore cannot stand. Then, with a more logical theory to take its place, it must soon become friendless and eventually die in obscurity. However easy to perform, this is a task I would rather shirk, and did at one time seriously think of omitting it from these pages since I have so many other drains on my vital forces, a limit of time and want of space. Finding the work imperfect without it, I thought it would be better to offer a few stray hints than to leave the subject wholly unmentioned, even if the arguments are not all I wished them to be, nor satisfactory to the general reader. Enough, however, will be given that the thinking mind may grasp the contents of the subject, while the student of nature will have an ample foundation on which to rear a greater structure if he wish to, or, for the critic to rend asunder the whole fabric of argument thus woven, and scatter its worthless fragments under the feet of a rejoicing multitude. At any rate, I will give enough of the theory for all practical purposes, and then leave it with the candid reader to dispose of as pleases him best.

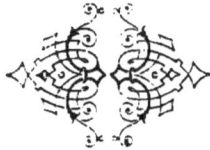

CHAPTER VI.

CRITICISM.

Without going back to inquire after the sun's origin, his orbit, physical constituents, or the cause of his motion, I will take him as he is, or was supposed to have been, when he possessed all the matter belonging to the solar system.

1. It is not necessary to ask why he began his rotary motion. It is sufficient to know that he revolves on his axis.

2. It is not necessary to ask why he became hot. It is all sufficient to know that he was in that state when he flew to pieces.

3. It is not necessary to ask why or how he became imbued with centripetal force. It is only necessary to know that he possessed attraction. Conceding all of the foregoing points to be non-essential, I will try, without their assistance, to show the fallacy of the alleged forces in producing the wonderful phenomena of planetary motions.

If the sun possessed all the functions accorded to him, we will have to presume that he honestly came into possession of them, and that, too, by a natural cause. It will also be necessary to presume that, like other bodies, he grew, and therefore must have increased by a natural process of growth. After reaching a certain size, the centrifugal force became too strong for the centripetal to hold the mass of matter thus collected, and therefore a part of it was released from the main bulk and hurled far into space. Admitting this to be true, the loss of a part of this huge bulk of matter would not change the original condition of the nucleus, and therefore it would continue to collect matter as before. In the course of time the accumulation of matter would again equal the former body, when the overplus would be cast off Thus would the process continue till all of the available gaseous matter had been transformed into solar matter and thrown off into space.

This is the only logical view to be taken of the subject, if indeed it has any logical side to it. Then, of course, the planets would not only all be of the same size, but would occupy the same position in space, because centripetal could hold together just so much matter against the power of centrifugal force. Therefore, when a given amount had been collected, another mass of matter would then be thrown off into space. If the masses were of of equal size, which they would necessarily be, since the two operating forces would be the same all the time, then, if they are equal in size, they would reach the same point in space, for bodies of equal size, projected by the same force, must necessarily reach the same point.

The sun now revolves on his axis once in 25 days, 14 hours and 8 minutes, and there is no reason to think that his rotary motion was different at any former time. Heavenly time tables always remain the same. No changes are made, so far as known, to accommodate theories, science or religion. But admitting that one mass might have been a trifle larger than another, does not account for the wide difference existing between the size of the respective orbits of the planets; therefore, their differences cannot be explained by that hypothesis.

Neptune and Uranus are computed to be equal in size; therefore, if projected by the same force, they would have reached the same point in space and now occupy the same orbit or orbits very near together. Yet, the former planet is said to be twice the distance from the sun as that of the latter. Saturn is computed to be one-sixth smaller than Jupiter, yet he is nearly twice Jupiter's distance from the sun. Mars is computed to be only one-sixth the size of the earth; therefore, if projected by the same force, he ought to have reached a point in space, far beyond the earth, if the ratio between the earth and himself was the same as that of Saturn and Uranus; but he is said to be only one-third farther from the sun than is the earth, while Mercury, which is the smallest of the principal planets, is the nearest to the sun; but admitting that the planets, by some inexplicable force, were detached from the sun and transported to their present places in space, all in good condition, still we find a difficulty in getting them trained into their proper places. The sun being a magnet, it necessarily follows that all

other suns possess a similar force of attraction. Therefore, while the solar orb exerts an attractive influence
over all other heavenly bodies, it necessarily follows
that they must exert a similar influence over him, and
consequently over every fractional part of him; which
being true, the planets would be influenced by them in
proportion to the size of matter each contains.

Let us suppose the earth to have been thrown
from the sun; the cause of it being detached from that
body was the inability of a centripetal force to sustain
that bulk of matter intact with the parental mass, and
therefore let it go, when the earth flew off 92 million
of miles from the sun.

Now arises the question: Could centripetal force
check the flight of a discarded mass of matter and
bring it to a circular motion, and hold it at a comparatively regular distance from the sun, while moving at the rate of 65,000 miles per hour? According
to physics, the ratio of force of projected bodies de-
creases with the square of a distance; therefore, as the
earth receded from the parental center, the sun would
lose instead of gain power over it, and, inversely, as bodies approach each other, the force of attraction would
increase with the square of the distance, which would
have a tendency to carry the earth in a straight line,
and eventually cause it to pass out of the solar
system. If centripetal force were unable to sustain that weight of matter when it was in the most
favorable position to be controlled, it is not likely that

it could call it to halt and bring it to in a circle about his own body, when so far away.

But the advocates of this time-worn theory no doubt will deny that the fixed stars have an influence over the matter within the solar system. In fact, they are compelled to do so in order to sustain their theory; but if it were possible for the sun's attraction to force the planets in a circular motion, it is not a logical conclusion that they would remain in exactly the same path for millions of years. Even if that were possible, all must agree that their orbits must describe a perfect circle, each degree of which should be an equal distance from the center of the sun; in fact, it could not be otherwise, if the sun produced all the force which controlled them, but, on the contrary, he is not in the center of the orbit of a single planet. If, as is alleged by the advocates of the Kepler theory, the solar system once had two suns, or centers around which the planets revolved, and which caused them to move in an elliptical path, then they would all elongate toward the same point in space, but since they do not, and no ashes or cinders of the missing sun were found, the agnostic will ask for an explanation.

The earth's orbit is a decided ellipse, while Mars varies from the true circle 26,868,C00 miles, about 1-6th the diameter of his orbit, which facts alone should destroy the old theories without any other evidence or explanation. But even if this motion could be intelligently explained, there is another that would puzzle

a couple of brilliant professors to elucidate to a
natural philosopher. The planets, in moving about
the sun, do not move in a regular curve, but execute a
serpentine movement; why they do has never been
satisfactorily explained by them, but if this, with all
of the other mysteries connected with the unreasonable
theory could be satisfactorily explained, the following
will forever silence the advocates of centripetal and
centrifugal forces.

CHAPTER VII.

PLANETARY MATTER.

If the sun was once a mass of molten matter and also a mighty magnet, by a gradual process of growth increased in size, he was then composed of a given combination of matter, because his nucleus would attract no other kind than its affinities. Therefore, all different masses of matter which were thrown from that body must necessarily be of the same chemical constituents, to which fact all must agree. Therefore, all of the planets thrown from the sun must be the same in their physical and chemical constituents. Venus has elements peculiar to herself; Mars has elements peculiar to himself; while Jupiter has elements peculiar to the nucleus which created him—which facts alone are enough to destroy the old theory in the estimation of all astrological students, even if others persist in clinging to it.

There is a vast difference in the size of the respective planets, which ought not to be the case. In the color of the lights they shed there is no resemblance existing between any two of them. In their diurnal motion there is but little resemblance existing between any two of them. In their influence in the human brain, which is a crowning evidence against the world-wide theory, there is absolutely no likeness whatever.

If all this evidence taken together would not destroy the groundless theory of the above-described solar forces, then human reason is hard to reach.

NEBULA THEORY.—The neblua theory, too, finds a place in the pages of philosophy, but it, too, is wanting in the essential points of a science, and will not gracefully bear criticism.

If the supposition is true that all the matter now belonging to the solar system was once collected in one vast molten body, of course it is spherical, since that is the form of all constantly moving bodies. As it condensed and cooled, according to theory, the center shrank from the surface, and thus cracked and discarded an outer shell. The shell thus discarded by the central sun fell to pieces, and was attracted together and formed a planet.

Of course, that theory is easy to understand, after it is known how many pieces the broken shell produced, and which way they slid off the sun, and which was the largest, and how fast each piece had to move to overtake the one that was moving as fast as they were, and what force detained them in their orbits while the sun shrank away from them, for he must still exert the power of attraction over them. No wonder the astronomers wanted to dissolve partnership with the astrologers, so they could find time to calculate the motion of matter and explain theories. But if the theory is correct, the first shell discarded should have produced the largest planet, and overcome the force of the sun's attraction, till all the pieces had overtaken each other and formed a planet. The second shell would produce a planet second in size and the third still smaller, and so on till the last one was reached, which would be the smallest of the entire set.

On the contrary, we find no such a regularity in their sizes and positions in space, the planet produced by the first shell discarded being only one-tenth the size of Saturn, though his shell occupied half the distance from his remote side to the remote side of Saturn, from the sun, and perhaps still more space than that.

Uranus, the planet produced by the next shell, contains the matter which occupied half of the remaining distance from Neptune to the sun and should not be nearly half so large as Neptune, yet he is computed to be equal to, if not greater than he.

Saturn comes next and the matter composing this planet occupied half of the remaining distance and should be still smaller than Uranus, but he is said to be one-sixth larger.

Mars comes next, and should be, according to the law of decrease, much larger than the earth, but on contrary he is six times smaller. Thus we find no evidence to support the nebula theory.

There are many nebula theories, but none of them satisfactory. In the foregoing theory, the planets should all be alike in their constituent elements, for the nucleus which attracted the chemical elements would attract but one combination, consequently each would produce the same colored lights and have the same influence in the human brain, which they do not.

DENSITY.—The density, of course, would have something to do with the size of the body, but even that throws no light on the subject, for it is a reasonable presumption that the first body thrown off would reach its greatest density first, since it would have to be cold enough to crack and break loose before it could be discarded. But according to the calculations, the last planet thrown off has the greatest degree of density.

ANOTHER THEORY.—If the sun was a hand-made machine, and manufactured out of molten matter, and set to spinning around like a musical top on an open floor, we can better understand how he slopped over and scattered his fragments throughout the solar system, regardless of size, form or regularity, before he settled down to a level-headed speed; but the theory would involve too many perplexing questios for practical purposes; for the small boy would ask who made the sun and what was he made of? Who melted the lead to make him? What did they mould him in and how long was he cooling? How long was the string they used to make him spin? Was it a big man that set him going, and could a small boy make him spin? What kind of a pavement did they start him on? and diverse questions calculated to distract the mind of his mamma and cause the Sunday school superintendent's head to get an extra dig, and make the school teacher cross for a whole week.

CHAPTER VIII.

PROPELLING FORCE.

Since having disposed of centripetal and centrifugal forces, it will be necessary to supply the planets with a propelling power, which is more forcible in its operation, more comprehensible in its explanations, and more logical in its conclusions, than the discarded theory embraces, or the tearing down arguments as previously set forth, will not be acceptable to the philosophical reader. According to the deposed forces, the sun is the absolute ruler of all force in motion existing within the solar system; but I shall attempt to show that he plays only a small role in the beautiful panorama of celestial motion. Elsewhere I have set forth theoretical explanation for the formation of the planets, and how they found their places; but their exact motions in circumscribing their orbits was not given. Consequently, I will now proceed to explain the cause for their very eccentric movements. All of these mysterious bodies move about the sun in perfect order and harmony, and are continually following the same unvarying courses which they have pursued since their first cycles were completed. So true are they in their motions that any point in the heavens which a given planet may occupy at any stated time in the near future may be calculated with a marked degree of precision. Owing to their uniform motions it is evident that there is an unabating force, which sustains the planets in their celestial rounds, or occasionally they would vary from their usual paths, even if they did not lose their bearings altogether.

We have been informed by astronomers that the planets disturb each other when moving near together. That being true, it would not be out of place to ask what force in nature settled them back in their places after being drawn or crowded out of their eternal pathway for a considerable length of time, while moving side by side, which some planets do for years.

Saturn and Uranus are at this time, 1895, very near together, where they will continue to be for years to come, without producing any commotion in the heavens or causing any unusual results so far as known. Of course they must interfere with each other now if they did when Dr. Herschel noticed a change in the movement of Saturn when Uranus was discovered. Owing to his great distance, it would be impossible for Dr. Herschel to be able to discover Saturn's eccentric movements, with his very imperfect glass, even if they occurred. If centripetal and centrifugal forces located the orbits of the planets in the way the theory describes, I am unable to understand how any planet can be pulled, pushed or jostled out of its accustomed orbit and yet remain unvarying in its motion, and be promptly on hand to make connection with all astronomical calculations, which are sometimes made years ahead of time; per-

haps the reader can, but those who are delicate must
receive it in doses to suit their mental digestion. The
author acknowledges his inability to successfully dispose of such indigestible philosophy. How a planet
could fly off at a tangent for a short period of time, or
even affect others to the extent of causing the slightest
deviation from their true courses, even when they are
moving side by side, I am unable to understand, and
especially is it improbable under the old theory of
centripetal and centrifugal forces. Since the solar
forces are insufficient, in the estimation of the author,
to create and sustain the many motions executed by
the planets, he feels called upon to replace the discarded theory with the following

ILLUSTRATION.—Though subject to revisions and
corrections, it is not wholly wanting in the essential
of a science. Whatever it may lack in detail, it has a
thoroughly scientific basis for its origin, and therefore
cannot fail to attract attention. It was discovered
after many years of hard and efficient study in connection with the zodiacal forces. These forces, for
thousands of years, have been known to affect the
human body. It is therefore not a visionary scheme,
concocted for the special purpose of deceiving the
reader, as shown in the following evidence. There are
two forces existing in the human brain called attraction and repulsion. They produce all likes, dislikes,
friendship, enmity, and assist in producing all of the
love and hatred that exists in the human family; but
which forces have not been clearly understood. Why

strangers at their first greeting form ties of lasting
friendship, and others fall passionately in love at the
first meeting of their eyes, has always been a deep
mystery to all save the interpreter of the language of
the stars. Why dislikes should enter the brains of
two entire strangers was hard to understand.

The principal influences which produce the above-
named effect are known to originate with the zodiacal
divisions of the heavens, and effect the different class
of people as follows: There are as many different
classes of people as there are divisions of the zodiac.
These twelve classes are sub-divided into many classes.
In the first class, which is ruled by the movable signs,
there exist but little sympathy, few likes and no love.
Among those born under the rule of the next class is
found much sympathy, friendship and love. After-
demonstrating the foregoing facts, I came to the conclusion that the human body could not be so radically
affected, or, rather, hold such strong affinities and an-
tipathies, for the zodiacal elements and the earth remain wholly unaffected by them. This thought led to
a further inquiry, and finally to the conclusion that
the movements of the earth were produced by zodiacal
and not solar forces; but I found some trouble in com-
paring and harmonizing the influences thus discovered, since the zodiacal forces were not just the same
as they were on the earth.

The earth and the moon I found to be attracted to
alternate zodiacal divisions, while the human body
appeared to be attracted very irregularly, which made

the influences quite difficult to understand. But since
it was known that each division of the heavens imparted a certain magnetic influence to all persons
born at its rising, which, without any opposition or
counteracting influence, would cause each person to
act on all others the same as the zodiacal influences
upon the earth. It was found upon closer examination that the effect was different. Then, in order to
trace out the zodiacal influences in man, and learn
how they would influence each other magnetically, I
selected twelve persons, each of whom was born at the
rising of a different zodiacal division. I then arranged
them in a circle corresponding with the zodiacal signs,
and then traced out their influences in each other, and
thereby learned their attracting and repelling forces,
as affecting each other.

Beginning at one born at the rising of Aries, I
learned that he was attracted to the one born at the
rising of the division of Taurus and Gemini, but repelled by the one born at the rising of Cancer. Then,
again, he was attracted to the one born at the rising
of the sign Leo, but with the one born at the rising of
the next sign, Virgo, there was a neutral influence observed. The next sign, Libra, I found, produced an
inharmonious feeling toward the Aries character.

Passing to the next sign, Scorpio, I found a neutral feeling existing between the two. The next one,
born at the rising of Sagittarius, was attracted to the
Aries person. The next man, born at the rising of
the sign Capricornus, was repelled, but the next two,

Aquarius and Pisces, were attracted to Aries. Thus I found the effect was the same, both right and left from Aries, till reaching the opposition sign, Libra. On each sign two are attracted, one repelled, one attracted and one neutral, when the opposition was reached, which produced a repulsive feeling. Then, taking the Taurus man and passing him around the circle, I found the order to be the same, but he was attracted to and repelled from different ones. Thus I found the Taurus man to be attracted to the first two to the right, which was Gemini and Cancer, and repelled from Leo, attracted to Virgo, neutral with Libra, and repelled by Scorpio. Then, returning to the starting point, and passing the other way, I found the attractive, repulsive and neutral feeling the same as in the former case: attracted to Aries and Pisces, repelled by Aquarius, attracted to Capricornus, neutral to Sagittarius and opposed by Scorpio. Gemini, like all the remaining signs, work the same. This man was attracted to the first two on the right, which was Cancer and Leo, and repelled by Virgo, attracted to Libra, neutral to Scorpio and repelled by Sagittarius. On the left I found the Gemini person to be attracted to Taurus and Aries, repelled by Pisces and attracted to Aquarius, neutral with Capricornus.

At first I thought their influence a little irregular, but on making the application I found the rule applicable to each zodiacal sign. But why the influence varied, as it did, with the alternate divisions of the zodiac, was a problem I found very difficult to comprehend. The only explanation I could find for such eccentricities in nature's work was that the earth, being somewhat foreign to the zodiac, had an inflence of her own which the zodiac could not altogether over-come, and that a child being born at a given point on the earth could not receive in full force all of the zodiacal elements, owing to the position he occupied, hence the eccentric influence is found to exist in the people. The earth, not having any environments, or rather being a part of the great economy of nature, had affinities in the alternate divisions; therefore was attracted to six and repelled from six of the zodiacal divisions, which causes her to exercise a serpentine movement. The moon being composed of similar elements, occupies the same orbit as the earth, but its nucleus being somewhat different, causes it also to be attracted to alternate signs, and the divisions of the zodiac which attract the earth repel the moon. It the nuclei of the earth and the moon had been the same they would have both occupied the same place in space, consequently there would have been only one instead of two bodies, or else one would have followed the other in the same path, but as it is they cross and recross each other's orbits in their journey about the sun. These attractive and repulsive influences of the zodiac cause the vibratory motion of the earh, as well the moon; but the zodiacal divisions which cause the earth and moon to vibrate may not affect the other planets just the same, though they all have a similar motion in that particular; but they may make a different number of vibrations in completing their orbital journey. Inasmuch as they are composed of a combination of elements the zodiacal affinites of Jupiter may be very different from those of the earth. The planets have another motion which produce an elongation of their orbits, which motion I also explain by zodiacal causes. Some of the planets have a stronger affinity for a zodiacal division than others do; consequently they are carried further in one direction, which cause their orbits to be elongated. If all the divisions of the zodiac possessed the same magnetic influence over the planets, they would all revolve in a perfect circle about the sun; but instead of that they each follow elliptical orbits. Stranger yet is the fact that no two of them elongate toward the same point in space, as the following cut will show:

The orbit of the planet Mercury elongates towards the division Sagittarius. Uranus elongates towards Aries, Saturn toward Capricornus, while the Earth elongates towards Scorpio. Thus it can be seen that none of the planets have a circular path. The explanation of these laws by the zodiac is the only explanation for the phenomena. I have not yet reached an explanation for the motion of the earth upon its axis, but I have lately developed a hypothesis which I will present to the reader in its imperfections, hoping to give it more time and attention in the not distant future, and have it ready for the next edition of Stellar Dust.

The magnetic needle is said to become unsettled when it reaches a point thirty miles south of the equator.

The cause of the peculiar behavior of the needle I attribute to the fact that being the magnetic center of the earth all of the zodiacal gases are received in that belt and pass into the earth, where approaching from opposite directions they meet and cause a whirl of forces which keeps the world in a regular motion on its axis; then passing to the north and south, they flow towards the poles, thus causing an electrical current to continually flow in those directions, and in escaping at the north they produce the northern lights. I am unable be account for the revolution of the earth upon its axis in any other way. Hence I offer the above hypothesis.

CHAPTER IX.

CONSTITUTIONAL LAWS.

The heavenly bodies are so situated, and chemically constructed, that they constitute two separate and distinct forces, which I will denominate constitutional and immutable laws, for the reason that one is paramount to the other, and that the former limits the operation of the latter in the execution of all their work in nature.

The constitutional laws are executed by the divisions of the zodiacal belt, and since they never change their relative positions to one another their individual effect upon the earth is always the same. However, an apparent change is sometimes produced, but which is caused by the intervention of the mutable or planetary aws. The twelve chapters of the constitutional laws, found in the great book of nature, are named as follows: Aries, Taurus, Gemini, Cancer, Leo, Virgo, Lbira, Scorpio, Sagittarius, Capricornus, Aquarius, and Pisces. Each of the foregoing divisions execute a separate and distinct law from the other.

The constitutional are the inexorable laws of nature from which there is no appeal. They produce all species of life that now is, ever was, or ever will be, on this globe; for life of any kind cannot exist independent of these mighty forces, and when they are once withdrawn then all life is doomed. They give size to body, form to man and species to all life. They also produce planetary motion and hold the world in its present position, and sustain their poles in their present positions relative to the north star and the southern cross. They also produce the zodiacal and equitorial aspects, the same today that they did at the beginning of the earth's existence, and have continued to perform the same duty since the cycles of earth's time began to run. They cause the earth to complete her diurnal revolution in 24 hours 48 min. 48 s., and speeds her in her course about the sun. They bring the joy of spring, the golden harvest of summer, the fruit of fall, and the hoary cloak of winter. They are the all-prevailing forces of the universe.

These laws doubtless originated long prior to the earth, or while it was in a gaseous state, and they have continued to rule and reign over it since that time down to the present moment, and moreover, they will continue to exercise that unrelenting authority over her long after man has any use for her support or protection. The earth will die by a slow process of decay. It will first lose its attraction for stellar gases in part, when they will slowly pass off into space nevermore to return. Then a loss of vitality will be followed by old age and sterility. In fact the process is imperceptibly going on at this time. The water is gradually departing from the earth; animals are becoming less hardy than former, while many species have become extinct, doubtless for the want o food containing a sufficiency of the necessary ingredients to give nourishment to their particular brains. This is the cause in part while the climatic changes have also had the effect to shorten life in the less hardy animals. The weaker go first, and finally others disappear. It is not a law of the survival of the fittest that protects and prolongs life in certain species. But it is the vital forces which impart to them strength and endurance, sufficient to withstand the climatic changes.

The carniverous will survive longer than the herbiverous animals, and will linger about the fishing places until the means of sustenance has disappeared from the face of the earth in the form of food and pure air. Then Byron's tableau will be on the scene: "Men will linger about their camp fires, the meager will by the meager be devoured." Except the water animals man will be perhaps the last to leave the dreary waste of lifeless matter which, like the moon, will swing its huge bulk through cold trackless space for an eternity of time, carrying with it the ruins of mighty nations, the crumbling thrones of once happy queens, the sepulcher of kings and cruel monarchs, and the white bones of departed greatness.

But, dear reader, don't be frightened for fear you will miss your Christmas dinner, because of all life coming suddenly to an end; for there is not the least danger of such an event taking place. If you meet with no accident, your health remains good, and you have a dollar to spare, you can have your Xmas dinner served in good form, even according to the Queen's taste and you can enjey it without fear of molestation, because of the wrath of nature reeking vengance on this cold, cruel earth. When it does come to an end it will be after a process of disintegration. In creating and destroying worlds, nature never gets in a hurry. She shakes up old mother earth occasionally to remind man of her mighty forces; but not because she is weary of supporting her children, and desires to shake them off, or to destroy the marvelous work of her creation.

CHAPTER X.

HOROSCOPE OF CHARLES DARWIN.

Mr. Darwin was born at the rising of Uranus in the sign Scorpio. When Mercury was casting a trine aspect to that planet, and also when the sun and Mars were in trine to each other. Uranus in Scorpio made Mr. Darwin an independent, original man, despising the beaten track of " beliefs." Mercury in trine to Uranus also made him original and intuitive, and gave him a brilliant, scientific imagination. The sun and Mars in friendly aspect to each other gave him great mental and physical endurance.

The foregoing influences made Mr. Darwin what he was in magnetic force and mental powers. He was a bold, progressive, original, intuitive man, such an one as is numbered by the sands of the centuries. But, his circumstances, and not his ability, gave him fame. Had he been born of humble parentage, he would have developed into one of the greatest cranks the world ever knew. There is but little in his horoscope, except his great ability to give him name and fame, and since his mind was carried off in such a strange direction, his progressive ideas would have been ridiculed and laughed to scorn. But, being backed up by an ample fortune to support him in all of his researches, which were many, arduous and tedi-

ous, he compelled attention. He was independent of
the world, and when he got his ideas in book-form,
they were found to be too searching to be laughed at,
too profound to be ignored, and too deep for the ordinary mind to grasp, which place this great man above
and beyond the reach of all. But, without the means
to help himself, he would have been considered a
dreamer or a wild, visionary schemer.

His mind did not run to money-getting; it was
scientific, deeply profound and penetrating to the extreme degree. But for wealth he would have lived in
obscurity, died in poverty, and his half-fledged ideas
would have fluttered out of existence at his departure;
but with it he rose to the crest of the billows of ap-
plause, and the world bowed to his desire, not wholly
because his theories were incontrovertible, but because
he was able to stand alone in the giant strength of
his manhood, in defiance of the opinions of the Christian world, in defence of what he considered to be
scientific and right. In poverty he would have done
the same thing, but to less purpose, for which reason
I place this man in the front ranks of the greatest
men the world ever produced.

CHAPTER XI.

ABEOGENESIS OR SPONTANEOUS PRODUCTION.

The point has now been reached which involves the question of spontaneous production, or as Huxley defines it, life out of not living matter, which is a misnomer, or how nature produced the first pair. All of the necessary conditions for the production of life having been explained, except sexes, which will receive attention further on.

First it will be necessary to locate nature's incubations which created the first life.

As the eastern hemisphere was, or is supposed by some, to have been inhabited before the western world was populated, it is presumably true that animal life began its existance in the old world, which, however, is only a presumption; for it is more than likely that life was created at every point around the entire globe at or near the same time.

On all islands of any extent have been found human forms, and each tribe differing in a small degree from the other. The difference, however, is more noticeable in the intellectual and facial than in the physical development, which is evidence that they did not spring from a common parent, but were produced under slightly different stellar forces, which would account for their intellectual difference.

The earth's surface having undergone so many changes since the first crust was formed, it is quite

impossible to tell just when life made its debut on the
earth, nor how long the conditions lasted necessary to
produce the higher forms of life. But presuming that
no radical change has taken place on the earth's surface since nature began to produce life, it will be safe
to locate the original birth-place of man on the East
India Islands, Moses to the contrary notwithstanding.
However, there is no special reason why life should
not have originated in, or near, the Mediterranean Sea,
as well as in mid-ocean, except that the position of
the former point is less favorable for the reception of
the zodiacal forces than is the latter place, it being
low, moist and well protected, fitted it for the require-
ments of nature in her work of producing life.
Owing to its suitability, it has been selected as the
most probable point on the earth if only one was
chosen for the beginning of life. Furthermore, it has
been located directly on the equator, and under the
ecliptic, where the electrical currents from the zodiacal belt play with more energy than at any other
point in higher or lower latitudes, which increases
its adaptability for the purpose.

Here originated the primordial germ, and here
vast fields of protoplasm were formed by nature's
subtle forces. Protoplasm, like other matter, appeared
on the earth by a gradual process of growth. Its beginning was the beginning of all physical life, regardless of what may be said of atomical life. It may be
doubtful logic to assert that there exists an atomical
life possessing intelligence, nevertheless there is some

reason for such a view. However, I would hesitate to affirm that the president of a club of atomical characters called a meeting and proposed to organize an animal compact for the purpose of furthering their individual interests, and thus created life.

But we cannot follow all the way down the scale of life, from man to the monad, or atom, and thus learn all of the eccentricities of the lowest possible form of life. Neither can we begin at the atoms and trace out all of their peculiar combinations up to man, and, therefore, cannot know all of the facts connected with their mysterious habits — their likes, dislikes and social and business customs.

It is not safe to say that man is the only intelligent animal on the earth, since it might be an unpleasant task to prove it. Then, if we accord intelligence to animals one step lower in the scale than man, it will be a difficult matter to stop at any point short of the atom. However, I can see nothing unreasonable in taking that view of the matter, since human intelligence is the result of a closer association of individual atoms of matter than when in their gaseous state.

I confess that it would be a difficult task to ascertain the degree of intelligence possessed by a single atom, or why one should possess more intelligence than another; but doubtless they do, as the result of their combinations will testify. Twenty, nor even fifty billions of inanimate atoms could not under any condition produce a live brain; but that number of live atoms could, under certain conditions, produce life and a degree of intelligence; but because a specific combination of atoms do manifest intelligence, it does not necessarily follow that all combinations of gases produce like results. The combination which enters into the construction of the brain is the only mind-producing matter man knows anything about; however, many more may exist.

Flesh, bone, wood and stone are composed of gases, but it can scarcely be understood that they possess intelligence; however, I have heard intelligent men defend the position that adhesion is a manifestation of intelligence, and however hard to prove, there is room for such a view, if the reasoning is carried to the extreme limit. The force which holds an iron plate together is claimed by the advocates of this idea to be an intelligent one, but when that plate is broken to pieces and ground to dust, then that force is apparently lost, and the supposed intelligence is not acute enough to reunite the scattered particles and hold them together as before without assistance, but if that assistance is rendered by reducing them to a molten condition, then the molecules of liquid matter, with an apparent intelligence will rush into each other's embrace; and, finally, when cooled, reunite as before in forming a solid iron plate, in which condition they will remain until separated by a force greater than their own.

When heated the molecules of liquid matter become active, and manifest an active intelligence, but when cooled, they can only manifest a passive intelligence, if it can be called intelligence at all. The clod of clay when dry is passive, but when it is moistened it manifests an active intelligence by dissolving. A ball of wet clay will shrink when drying, but why the molecules will cling to each other, and move toward the center and crowd closer together, is not clearly known.

By this process of reasoning it is possible to grant intelligence to all matter, but I prefer to draw the line of distinction between animate and inanimate intelligence in all physical matter.

Vegetable life manifests a greater degree of intel' ligence, than does the mineral kingdom, but I much prefer to recognize the force in the mineral kingdom as attraction undefined, and that of vegetable instinct, for I think the vegetation acts many times instinctively but not intelligibly—flowers following the sun; the interchange of pollen; the growth of one plant on another, as the licorice root, or plant instinctively grows on the limbs of forest trees, and accord to all animal life however minute in form intelligence undefined because there are many degrees of each which require special definition. But these are unimportant technicalities which I shall not discuss just here.

But brain life and human intelligence with its many degrees of power are before us, but the exact modus operandi pursued by nature in organizing matter into intellectual forces may never be known for the reason that the atoms themselves of which the respective brains are composed are too unfamiliar with man to be fully comprehended by him. It is plain, however, that nature executes her work by force, and that all of her products are ingeniously, systematically and scientifically produced. If she works by force, there must be a law back of it to impart energy to matter and thus give it executive power.

However wise and learnedly men may talk of natural laws, they must necessarily have a very poor conception of them until they are able to define them; at least they must presume to understand them and be able to define them, according to their presumption. To say that a law exists, without being able to define it, is talking at random, even if the law does exist, since no one can intelligently converse on a natural law, or any other subject, without first understanding it. So, before going further in the discussion of this question, I will define natural law as I understand it to be, and leave it to the reader for his candid consideration. But before definining natural law, I will draw a comparison between it and civil law to show their similarity.

CIVIL LAW.—The printed code is only an expression of organized force, existing in the people and it shows to what extent that force may be executed in maintaining justice and order in society by those delegated to carry it into effect.

The highest officers receive their executive powers from the people. The subordinate officers usually receive their instructfon from their superior officers, notwithstanding the fact that both are responsible to the people whom they serve. Thus, in civil law we have both principal and subordinate officers. Officers with their prescribed duties before them are powerless to act, as officers since they must conform to the letter of the law. Law may exist and the officers may be clothed with full power to act, but a stranger unfamiliar with the custom of the country might be present and not know that either existed, for until an individual oversteps the bounds of his legal rights, the officers cannot interfere with him. Therefore, both law and officers must remain as though they did not exist.

Again, an officer might be in a country, in which he had no jurisdiction; in that case he would be powerless to act in the capacity of an officer. Thus civil law prescribes the legal rights of a citizen regardless of its source, whether derived from a republic, monarchal, or tribal government, for in all cases the law resides in the people and not in the books.

NATURAL LAW.— Natural law is not altogether dissimilar to civil law, though civil law has nothing to do with natural law. But it is amusing to note how readily and perfectly man conforms to the force of natural laws in every department of life. Civil law, so far as can be known, is framed in exact conformity to natural law. In nature we find both law and officers the same as in civil governments. The fountain or forge of force whence originated natural law, is unknown. It is too subtle and remote for man's mental grasp, consequently it cannot be fully understood. Not being able to comprehend the great first cause, we are compelled to accept the second and third as agents

or officers of the first, and deal with them according
to the light we have. Nature is a wonderful organizer. She has all of her atomical forces thoroughly
under the control of the heavenly bodies, which are
the superior officers, and which receive their forces or
authority from the great first cause. The gases are
the subordinate officers and are directed in their work
by the heavenly bodies, though both, like civil officers,
receive their power from the first cause. Without
organization there could be no life, since individual
atoms of themselves have no executive power or intellectual selection, but are as much under the control of
the heavenly bodies as a soldier is under the control
of his commanding officer. Each atom works in the
ranks assigned to it, and is ever ready to execute its
work when authorized so to do, or when conditions are
favorable for its operation; but, like a civil officer, it
may not always have jurisdiction. In that case they
must remain in a passive state, in which condition
their existence might not be known. The proof is
seen in the following facts:

The laws which created the mammoth, mastodon,
and the æpyornis, the bird that laid an egg three feet
long, exist the same today as they were thousands of
years ago; but those quadrupeds and fowls have disappeared from the face of the earth, and no new ones
are being created, for the laws which produced them
are in a passive state, because the earth is cool, and
not because the laws themselves have changed. They
simply have no jurisdiction.

**The creative laws then are the unknown, but
organized forces of nature, innate in atoms of
matter, but directed in the execution of their
work by the earth and the celestial bodies.**

CHAPTER XII.

INCUBATING.

Before proceeding further with the explanation of spontaneous generation, or nature's mode of production, perhaps it will be just as well to review the patent process of incubation, which was nature's original plan of production, only conducted on a somewhat larger scale, which will make the subject of spontaneous generation plainer to many, if not clearer to all of the readers of this volume.

But before incubating, let us inquire into the nature of the thing to be incubated. We see the eggs go into the mysterious box, there to remain a given period, during which time we watch them closely, and occasionally examine one, and wait till life appears in another. Still nothing is learned of how the atoms were induced to unite in producing the animated thing. Dr. Dalton, in his famous work on physiology, explains, so far as he could, the chemical constituents of the egg. He informed his readers that it is composed of oxygen, hydrogen, carbon and nitrogen, and that during incubation it takes on more oxygen and throws off carbonic gas. The egg absorbs nearly two per cent of its own weight of oxygen, while the quantity of carbonic acid, thrown off at the same time, amounts to no less than 24 grains, the process going on during incu-

bation.   He further adds that it is the same in eggs as
breathing is in animals.

The foregoing explanation being true, plainly
shows that the animating elements necessary to produce life come from the outside of the shell, which is
further evidenced by the fact that eggs oiled before going into the incubator will not hatch, because the
pores of the shell are closed, and thus exclude the
necessary gases from entering the egg. But the
learned doctor leaves his students in the shadows concerning the chemical construction of the egg, or how
it proceeds to organize life.   From a chemical analysis
it cannot be learned whether the egg in question was
that of a duck, turkey or chicken.   The chemists can
find no beak-forms nor foot-prints on the yellow island
floating in that chrystalline sea to reveal the form or
character of its progenitors.   Neither can there be
found anything which will reveal the spark of life that
is supposed to slumber in that masked substance,
nor explain the cause of its differentiation.   To the
chemist the egg is a deep mystery.

Then, if from a chemical analysis, little can be
learned of the causes which produced the life within
the egg, nor lead to the discovery of its hidden mysteries, it will be necessary to proceed by mental
analysis in order to reach any definite conclusion concerning the matter; for reasons must reign where
demonstrations fail.

In order to learn the process of incubation it is
necessary to understand the organization of its chem-

ical constituents; for the egg is an organized body; but it contains no life within its walls any more than unignited wood contains fire; but the elements are there which under proper conditions will produce life in the one or fire in the other. The egg is only a bit of protoplasm, not altogether unlike the jelly-like substance which produced the original progenitors; the chemical elements of which the egg is composed are uot a homogeneous mixture, as they appear to be, but they are perfectly organized in every part, and possessing a given number of nuclei, but which are too subtle in their relations to each other for their individuality to be discovered by any process of analysis yet known to chemists; nevertheless, they exist.

Now, brother scientists, don't dispute that fact; for you know that the brain is composed of chemical divisions, called phrenological organs; yet many scientific men dispute phrenology because they cannot prove the mental divisions of the brain by chemical analysis. Nevertheless, the most ignorant as well as the most learned can testify to the truth of phrenology, notwithstanding Dr. Dalton says that the brain is a unit. Though we have no reason for denying that different parts of the brain may be occupied by different intellectual faculties, there is no direct evidence which would show this to be the case. The layers of gray matter in each principal portion of the brain is continuous throughout. There is no anatomical division or limit between the different parts, like those between the different ganglia in the other portions of the nervous system. Consequently such divisions of the cerebrum and cerebellum must be altogether arbitrary in character and not dependent on any anatomical basis. If the physical divisions are not observable in the brain, they are on the skull, and the chemical action of the respective chemical divisions produced them. Then it illy becomes a chemist or any man of science to dispute a subtle point in chemistry, when a glaring fact of like nature defies his skill. The egg contains three distinct physical parts as seen by the eye, which are, first, the cicatricula, the little white speck which is always seen in eggs, and which is the germ corresponding to the ovum of the mammal. Second, the yellow center. Third, the white albumen substance which surrounds it. In their chemical structure these parts in no wise differ from the orignal protoplasm. The egg is a physical growth, containing no active or visible life. The cicatricula is the germ, and contains all of the necessary elements for quickening. The other parts are the food necessary to sustain the germ in the process of its growth. The egg thus organized will produce life.

If the egg is an organized body some force organized it, and that is the agent scientists have long been looking for. The following is my solution for the problem. The brain of the fowl, like all other brains, is composed of chemical divisions, or brain centers, each of which snpplies its part of the life essence to the egg, while it is forming in the fowl. Thus each brain center creates in the egg a chemical nucleus corresponding to its own center. Each nucleus thus created in the egg is a center and magnet of attraction, and is complete and perfect in its chemical structure, after the male forces are received; after which they are locked up in the shell to protect them for future use. The egg receives its incubating forces from the brain of the fowl, and the brain receives its life essence directly from the zodiacal divisfon of the heavens in a pure state, by the process of inhalation and attraction. They are inhaled into the lungs when the blood receives what it needs and carries it to the brain, when the brain attracts its affinities and utilizes them in creating a life essence. The brain will not attract from the blood what it does not need for that purpose, therefore the blood will not attract from the atmospheric air in the lungs that which the brain does not need, therefore it can not impart to the egg elements foreign to its chemical divisions. Thus each brain center creates in the egg a chemical nucleus corresponding to its own center. Each division thus created in the germ is a magnet, which as soon as the necessary conditions are supplied, is ready to begin the work of organizing its part of the animal body.

The brain of the male operates the same as that of the female, in producing his fecundating forces; when the egg receives the male principle, every center in it becomes fertilized. Thus it can be seen that the egg can not possess a greater nor a less number of centers than is possessed by the brain of the fowl which produced it; neither can the male bird produce fecundating forces to accommodate a greater nor a less number of centers than those of the fowl belonging to his own species, for his elements must correspond identically with those of the egg, or they will not unite.

If from any cause either of the divisions of the female brain fails to supply a sufficiency of its forces to perfect its nucleus in the egg, that division will not receive the male elements, consequently the egg will not incubate. The male principle, too, might be defective in some particular, which would destroy its effect, therefore the egg would not fertilize when coming in contact with the spermatazoa, consequently it would not attract stellar gases; but if the egg is perfect in all its parts then it will hatch.

The brains of fowls belonging to different species slightly vary in their chemical structure, consequently in their chemical affinities, and, therefore, will only receive the combination of gases peculiar to themselves, which alone will produce forms characteristic of their own.

The turkey has a head adorned to correspond with its genus; consequently it must have a brain capable of throwing out a nerve branch to create its head gear, and also to produce in its egg the chemical centers necessary to produce the brain. What is true of one is true of all other fowls.

Now we understand the nature of the thing to be incubated. It is, therefore, in order to explain the process of its development.

PROCESS.—At one time it was generally supposed that eggs must be covered by a fowl, and thus receive animal heat, in order to induce them to hatch. But science laid that idea gently aside by demonstrating with a heated oven the fact that animal heat is not necessary to produce incubation. The egg in its perfect form is now placed into the incubator properly prepared to receive it. The heat warms it through, when the process of attraction begins and development commences where it left off.

The the nuclei in the egg was formed in the fowl, by the zodiacal laws, and now the same forces continue their work of development. As the world revolved on its axis, each nucleus attracted in their regular order, until the brain was formed. One branch developes the heart, another the lungs, and still another the kidneys, and so on, till the entire organism is perfected, at which juncture the magnets in the egg become consumed and must be supplied otherwise, in order to continue the growth.

At this point of development the chick grows hungry and, in pecking for food, breaks through its shell. The oxygen rushes in and invigorates the new life, when it struggles to free itself from its covering, to begin a new form of existence. It now takes food, and new magnets are formed in the blood, which attract their stellar affinities, and life continues. Nerves, thrown out from the brain, form every part of the body and develop every pecularity of the fowl. The brain of the chick sends out nerve branches corresponding to the number of nuclei it possesses. The duck sends out nerve branches corresponding to the

nuclei it posseses. The former has five toes and a
round beak. The latter has web-feet and a flat
beak. The duck does not have nerves in its beak because the beak is flat. but because the nerves grew and
formed it that way. For the same reason it has webs
between its toes. Species will not cross; therefore,
their differences must alway remain the same.

CHAPTER XII.

PROTOPLASM.

I will now explain the process of creating physical life, which began with the protoplasm.

First, let the reader get the right understanding of its nature, before he tries to comprehend the results of its growth. Then the whole lesson will be clear and comprehensible. Protoplasm, scientific men tell us, is composed of oxygen, hydrogen, carbon and nitrogen; but I am of the opinion that it contains many more than four elements. In fact, I know that the original protoplasm did, because life came from the protoplasm, and human life contains many more than four elements. When manufactured in the laboratory of the chemist it may contain only four gases, but perfectly developed protoplasm, capable of producing human life, contains no less number than twelve different chemical elements. The egg of the fowl is known to contain ten or more elements. Protoplasm from which evolved the human body, could not contain a less number of gases than the goose egg and produce a higher form of life than the goose.

In the National Museum there are on exhibition vessels containing the following named ingredients which are found in the human body of the average man, who weighs 154 lbs. A large glass jar holds the ninety-six pounds of water which his body contains. In other receptacles are nine pounds of white of egg, a little less than ten pounds of pure glue, thirty-four and one-half pounds of fat, eight and one-quarter pounds of phosphate of lime, one pound of carbonate of lime, three ounces of sugar and starch, seven ounces of fluorine of calcium, six ounces of phosphate of magnesia and a little ordinary table salt.

Divided up into his primary chemical elements the same man is found to contain ninety-seven pounds of oxygen, enough to take up, under ordinary atmospheric pressure, the space of a room ten feet long, ten feet wide and ten feet high. His body also holds fifteen pounds of hydrogen, which, under the same conditions, would occupy somewhat more than two such rooms as that described. To these must be added three pounds and thirteen ounces of nitrogen. The carbon in the corpus of the individual referred to is represented by a foot cubic of coal. It ought to be a diamond of the same size, because that stone is pure carbon, but the National Museum has not such a one in its possession. A row of bottles contain the other elements going to make up a man. These are four ounces of chlorine, three and one-half ounces of fluorine, eight ounces of phosphorus, three and one-half ounces of brimstone, two and one-half ounces of sodium, two and one-half ounces of potassium, one-tenth of an ounce of iron, two ounces of magnesium and three pounds and thirteen ounces of calcium.

Calcium, at present market rates, is worth $300 an ounce, so that the amount of it contained in one ordin-

ary human body, has a money value of $18,300. Few
of our fellow-citizens realize that they are worth so
much intrinsically. What makes this metal so costly,
is the difficulty of separating it from the elements with
which it is found combined in nature.

It seems odd to know that four of the constituents
of the human body will take fire by spontaneous com-
bustion. Everybody knows how quickly phosphorus
will do that when dry. A scrap of sodium, on being
thrown into hot water or upon ice, will burst into a
rosy flame. Potassium acts similarily; but with
greater violence. On touching water it flames up, and
at length explodes, throwing a mountain of sparks into the air. Magnesium, which is used in the form of
powder for flashlights by photographers, is so readily
and fiercely combustible that it has to be kept tightly
corked in bottles.

The growth of the protoplasm is necessary in
order to accumulate the required combination of elements to produce life; for until the necessary number
in sufficient quantities are gathered together in a single
body, nature cannot differentiate any form of life;
hence the necessity for protoplasm.

It is generally supposed that protoplasm grew
from its own volition, and that from that growth
sprang life; but the reverse is the case. It would be
just as logical to say the placenta was produced that
the child might be created, as to say the protoplasm
was produced prior to the germ. The germ of life
was formed first, that the protoplasm might be pro-

duced. As the vital parts of the germ were gathering their forces, preparatory to forming organized life, much refuse, crude matter necessarily formed about the germ, to protect it during its growth; hence the production of the protoplasm.

Protoplasm, like all physical matter, grew from a small beginning, and of course made considerable growth before taking any definite form or manifesting any active life, since that was and is nature's method of producing all forms of life. There must be a certain amount of growth before protoplasm can receive and retain a sufficient quantity of stellar forces to create the object of its growth, however small that object may be. Show to a man who has never seen a walnut, pear, or an ear of corn, and ask him how they came into existence. If he knew nothing about them he would be at a loss to give any explanation for their growth. If he was from a country that produced only toad-stools, potatoes and pumpkins, he would very naturally conclude that the ear of corn was pushed up through the soil, like a toad-stool, or grew on a vine, like a pumpkin. If, however, he was a philosopher he would know that a vegetable growth of some kind was produced before the corn could grow; but he could not describe the form of the growth, the size of the plant, nor the texture of the stalk.

In the animal kingdom the same law exists. A growth of some kind was necessary before animal life could be developed. It came in the form of protoplasm. It grew a formless thing till it reached a cer-

tain degree of development, when differentiation took
place. But now we want to know the character of the
life, or the degree of intelligence atoms of matter possess under the necessary conditions, to prompt them
to unite in producing the original protoplasm.

EXPLANATION.—Atoms, like the people which they
create, possess strange characteristics, for they will
not all, when under apparently the same conditions,
unite in the same way to produce the same effect.
Those gases only which possess affinities for each
other will affiliate; but whether the virtue of the association lies in the peculiar form of the atoms composing the body in question, in the affinity of the
atoms themselves (if any one knows what that means),
or in the degree of intelligence they possess, is yet to
be learned. But the latter I maintain to be true.
Nevertheless, that intelligence must be directed by
some other than the innate force, which atoms possess,
before they will embrace each other in producing
growth; and that director or directors are no other
than the zodiacal and planetary bodies. Protoplasm
was not a divine conception, but a mundane production and manufactured out of STELLAR DUST, and not
mundane dust. These gases were induced to unite by
the combined efforts of three agents of the earth, as follows: 1st, heat; 2nd, moisture; 3rd, motion.

Doubtless it never occurred to the reader that the
motion of the earth on its axis ever had or now has
anything to do with producing and sustaining life;
yet, that is one of her most important functions in

that particular, since the brain can receive gases only when in certain positions relating to division whence they come. Then, in order to occupy all of the positions, the rotation of the earth on its axis is necessary. If the earth had no diurnal motion, protoplasm could not have formed. Consequently higher forms of life could not have been produced originally; neither could they have been reproduced subsequently; nor could life now exist on the globe. Should the earth stop now at high noon, when the sun, the supposed giver of life, is in the mid-heaven in all his glory, all life would soon become extinct; the embrio would die in its mother's womb. even if the mother herself should longer survive, which is not probable, for the reason that the adult depends as much on the zodiacal gases for life as does the embrionic form. Therefore, the motion of the earth on its axis is necessary to the existence of all life. Moisture was necessary to prepare the zodiacal gases for their union, for gases will not unite in a growth of any sort without it; while the motion of the earth on its axis was necessary to bring any given point on the earth's surface in contact with each and every degree of the zodiacal belt once every twenty-four hours, in order to complete the conditions.

The perfect protoplasm to all appearances, is a homogeneous mass of matter; but nevertheless, it was as perfectly organized in its chemical structure as is the egg in the shell. On receiving the necessary impulse it grew apace, and finally took form, which afterwards became animated. It was just as easy for

nature to produce animal life after protoplasm was
created as it now is for her to produce life from the egg
after it has been laid. Moreover it was just as com-
patible with nature's laws to produce protoplasm when
the earth was at its proper degree of temperature for
that purpose as it now is for her to produce an egg in
the fowl. .It is conceded by all that the fowl produces
the egg within her own organism independently of
nature's external laws, but she does not. since she has
no power to reproduce apart from them. She only
perpetuates the conditions which the earth possessed at
the time of its original ancestors. The egg is only a
protoplasm produced by nature through the agency of
the fowl, and not by the fowl herself. The nucleus ..
which forms the brain of the original progenitors
is retained by all fowls of like species, subsequently reproduced, which enabled nature to reproduce similar nuclei in the egg of any given fowl. The
brain of the fowl, like all other brains, is composed of
chemical divisions or brain centers. and each center
supplies its essence to the egg.

LIFE FORCES.—As previously explained, each 30
degrees of the Zodiac throws out the combination o f
gases peculiar to itself and different from the other
eleven divisions of that belt, and which form an attraction to the oarth. Furthermore, each one of the
twelve divisions create a separate chemical division in
the brain. It further appears that each degree of the
specific division supplies a slightly different combination of gases from the other degrees from the

same division; therefore, when a nucleus begins to form
out of the elements belonging to any degree of any
specific sign; that degree determines the character of
the nucleus of the brain subsequently formed, because
it determines the combination of gases demanded
from the other Zodiacal divisions in producing
life. It having affinities peculiar to itself requires
certain other elements in order to complete its work.
Each degree of the zodiac is different; therefore, all
the elements from the other degrees will not affiliate
with any given one.

The union of the first two atoms which unite to
form the first epinucleus was the beginning. This
epinucleus holds the ruling position in the completed nuclius of the brain, which now resides in the
protoplasm. Thus the nucleus began at the rising of
the twentieth degree of Gemini, then the epinucleus
created from the degree of that division of the Zodiac
would hold a commanding influence in the nucleus,
and thus control the organization of all the elements
subsequently added to it. Should a nucleus form at
the 20th degree of Leo, or any other degree of that or
any other division, the brain thus formed would differ
from all other brains produced from other degrees of the
zodiacal divisions, because each degree has a different
affinity from all others, and therefore, would attract
other combination of elements to it. The first nucleus
is now formed and gathers as it goes As the world revolves on its axis this epinucleus is carried through
the gases from the next division of the heavens—which

is Cancer. During the time the point on the earth at
which the protoplasm is now being formed, is passing
through this division, a second epinucleus is being
formed and becomes attracted to the first; as these two
are being carried through the elements of the succeeding division. Leo, another epinuclei is formed and
becomes associated with the former one. Again, while
the three are being carried through the elements from
the next division the fourth nucleus is formed and finds
its place in the association of the other centers and
thus the process went on; till the earth had made one
diurnal revolution, and that point on its surface had
passed through all of the elements from each and all
of the twelve zodical divisions, and each had completed its epinucleus, and all of them were united to complete the perfect working nucleus of life which is composed of twelve minute chemical brain centers.

The process required twenty-four hours for nature
to complete her work thus far. Since it requires that
length of time for the earth to complete one diurnal
revolution. During the next twenty-four hours, while
the earth was revolving, and carrying the nuclius thus
formed through the same forces, more off them would
added to each nuclei, and thus growth continued. This
process was daily repeated until a complete protoplasm
or egg was formed and fructified. How long it required for nature to produce the egg is not known; but at
this point new forces entered inte the protoplasm where
embriotic life began. Still the process of growth
was continued, for the embrio thus formed could

not further developed without being supplied with
the same combination of zodiacal gases as the original. As soon as sufficient brain force was concentrated
each brain center threw out a branch of the nervous
system, each of which has an individuality of its
own, yet, they all worked harmoniously together,
in producing the body. The cranial nerves are named
as follows:

1st. Olfactory Nerves serve to convey the special
sense of smell to the brain.

2nd. Optic Nerves produce the eyes, convey
vibrations, blight and impart the sense of sight.

3rd. Auditory supplies vitality to the ear drum
and conveys the sound vibration to the brain.

4th. Motor Occuli helps to control the eyes, teeth,
jaw and tongue.

5th. Patheticus assists to supply the brain force
to the muscles of the eyes.

6th. Motor Exterminis also performs an oracular
function.

7th. The Fifth pair performs a facial function, and
control in part, the jaw, lips and nasal organs.

8th. Glosso Pharyngeal assists in controling the
tongue and mouth.

9th. Facial Nerves control the action of the face
and ears.

10th. Pneumogastric supports the heart, stomach
and liver.

11th. Spinal Accessory control the muscles which
rules the actions of the lung when talking or straining.

12th. Hypo-glossal assists in controling the tongue.

As these and the spinal nerves advanced, bones,
cords, flesh, and all the internal and vital organs were
produced.

The heart was formed first to regulate the flow of the blood as soon as it became necessary, for the heart is not a force-pump to produce circulation ; but a regulator to govern it's flow; for circulation began before the heart was formed.

The blood is composed of atoms which possess force, and so long as new forces are being taken into the blood by inhalation the blood will flow, for force and motion are innate in matter. At the expiration of nine months the life had so far advanced in its development that it required more nourishment than nature could supply through the circulatory system of its crude mother, therefore it was freed from its proto-plastic parents, to survive or perish as the stars of fate might determine; but as nutricious food was easily obtained in those very productive times, the stars of fate favored them; for life of all kinds sprang spontaneously into existence, upon which the early discoverers of the earth rapaciously fed, flourished and became mighty. They were not so particular then as now, in their bread and butter regime. They had not yet learned to demand broiled beef-steak for breakfast, a morning cocktail for an appetizer, four courses for dinner, and oysters on the half-shell for supper. For breakfast they probably took mush straight, prepared in the luke-warm elements in the murky marsh. Neither were they so fastidious then in their table etiquette as their fine haired descend-ents are now. Fingers were made before forks, and doubtless were used to fish worms out of the mud, and to catch the juicy bug on which to feast before they discovered beef-steak and the methods of preparing it, and the upper and nether stone for grinding grain, as well as the methods of mixing and baking the hoe and the Johnny cake, the unleavened bread and the light loaf, the soda and Carolina biscuits, the French, German and the hot rolls; the waffle, buckwheat cakes, fritters, Yankee dough-nuts, etc., etc.

SPECIES.—Human life now requires ten moons for the zodiacal brain to develop; therefore, it is presumably true that it requires exactly that length of time for the protoplasm, after it was fructified, to develop human form.

Whether the species date from the beginning of the nucleus or not till after its growth and fructification, by other elements not incorporated in the original protoplasm, is yet unknown; but I am persuaded to think that the moment the nuclei of the protoplasm is completed by the first revolution of the earth on its axis, the specie is determined; for there appears to be no reason for a germ thus begun to change its organization, nor its original combination of elements, since the conditions were ever favorable for the continuation of its growth since it had begun, for it could receive, without interruption, gases from the same source of supply as that which originated the twelve epinuclei. In fact they could not change, as shown in the development of the egg of a fowl in which the growth is arrested at the time it is laid, and it may not be resumed for weeks, yet no change occurs in its attrac-

tive power, nor the result of its incubation, because
the conditions favorable to its further development
have, for the time being, ceased, does not change its
magnetic force nor cause it to lose its power of attraction for zodiacal gases, for as soon as the necessary
conditions by the incubator are restored the process of
development is resumed, and goes on the same as
though it had not been arrested. This question might
now arise: why are the conditions not changed by the
cessation of the development in the egg if nature does
the work?

The answer is easy so far as species are concerned,
for the same zodiacal laws would take up their work
of development just where they left off, and proceed
with as much precision in the process of evolution as
an old lady would when she resumed her knitting
where she left off a day or week previous.

The ruling epinucleus would still preside over the
development of the chemical center. Then, when heat
was applied, and the earth had reached a point at
which it could receive the required elements which
most likely would be at the degree of rising when the
egg was perfectly formed before it was laid. Here
nature would resume her work and proceed as though
she had not been interrupted, for which reason I must
conclude that species began with the nucleus of the
protoplasm. If it did not then the protoplasm might
produce any species of life. On the same principle an
egg might produce any species of the feathered flock
by nature changing the chemical conditions of the egg

by adding or withholding at any time one or more of
the chemical elements belonging to it, which being
true, the species of the fowl would not depend on the
character of the egg, but upon the chemical constituents of the spermatozao of the male, which supplies the fortifying forces. If the egg was a homogeneous mass of chemical elements the male of any fowl
could fertilize the egg of any other fowl, and then
reproduce his own or perhaps any other species; therefore the common hen might produce a guinea, a goslin,
or a grouse, and a goose might produce a canary bird
or a pewee, a crow or a pelican, or a new species.
Consequently, no law or order could be maintained in
the production of life; but as the laws now exist the
epinucleus which are first formed in the egg, as previously described, determines just what the species
shall be and plainly show why they continue the same
and can not be changed.

The foregoing rules will hold good in regard to form,
but not in regard to intelligence, for the simple reason
that the zodiacal laws rule the development of the
physical form or species, and, since they never change
their relative positions to one another, they may suspend at the laying of the egg, and again resume their
work of development at any time within a given limit
with no inconvenience, delay or confusion of forces.
But since the planets rule the growth of the brain,
which produces the intellectual faculties, their effect
is different at different times, since they are continually changing their relative positions and aspects to

one another; and, as the planetary brain is produced
after birth, it does make a difference when a chick is
hatched or a child is born, and the latter in particular.
That the epinucleus of a protoplasm, which formed at
the rising of a given degree of the zodiacal belt, produced animal form may leave in the mind of thoughtful readers the impression that only 365 species of
animal life could be produced, since that is the number of degrees in the zodiacal belt. It will be necessary
to explain further.

The reader must remember that the earth's orbit
varies a little in performing its cycle, which requires
29 years. This variation from the exact latitude and
declination relative to other planets, might change the
combination of elements which composed the first
epinucleus, and, consequently, the other eleven, and
thus produce different species, even under the same
longitudinal degree of the zodiac; which being true,
would account for the multiplicity of species found
on the earth even if there was no other explanation
to be given; but, since the temperature of the earth
was constantly changing, species may be accounted for
otherwise. Different degrees of temperature might,
and doubtless did, cause different combinations of
gases to unite in forming the various species, since
the combinations are not all affected alike by the same
degree of temperature.

The nucleus formed when the earth was at its
greatest degree of heat that would produce life.
Nature created the largest and most tender animals,

many of which could not survive the chemical and
climatic changes which the earth has since passed
through, hence became extinct. As the temperature
of the earth ran down during the process of cooling,
different effects on the gases were produced, so that
many and varied were the nucleus formed before it
reached a degree that would produce nothing higher
than the insect life. The life-producing period of the
world may have covered millions of years, and millions
of species may have been produced, which were unable
to reproduce; consequently, their species were lost at
their death. Strong evidence in favor of this apparent
fact is that there are barren animals among all species;
then, if in reproducing nature fails to perfect one of
her species, she may have failed to have perfected some
of her original creatures. Besides, thousands of species
may have been produced, lived for a time, and finally
became extinct without leaving any trace of their
former existence.

A very natural supposition is that the interference
of the sun's heat would have made the earth too hot
to incubate any species of life; but this point is explained when it is remembered that 108 degrees is
incubating heat, and that when the earth was at that
degree of temperature water would evaporate very
rapidly, and thus keep dense clouds in the air, for the
clouds would have to become very dense before they
would condense for the want of cold currents, which
could not exist when the whole surface of the earth
was hot. These clouds would obstruct the sun's rays,

and prevent them from reaching the earth; but as time rolled on heat grew less, and cold currents were created; clouds condensed and rain fell in torrents. Thus were the clouds dispelled, while the sun was permitted to pour his violent rays with full force upon the earth which made the days intensely hot and the nights cool, which had the effect to bring the period of life-producing to a close, after which all animal life must be perpetuated by the sex-process, and incubation for mother earth had passed her change.

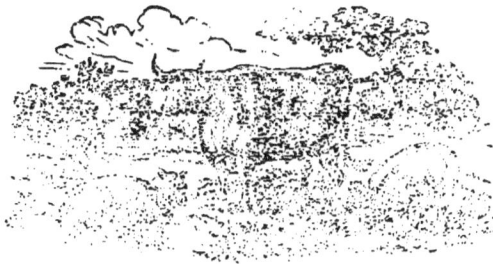

CHAPTER XIV.

REPRODUCTION.

In the reproduction of animal life nature works just the same in the human organism as she did when the earth alone supplied the conditions of heat and moisture. Though her apparatus was more crude in all of its appointments, nevertheless results were just the same.

The beginning of animal life in the animal organism is the union of the first two atoms of matter in forming the first epinucleus of the ovum, for the human ovum is made up of twelve chemical centers, just the same as was the first protoplasm. The ovum represents the germ of the original protoplasm, and is produced by the same zodiacal laws in the same manner as the first protoplasm was organized and assisted by the same conditions only supplied through different channels. The earth supplied the heat then; the mother supplies the necessary temperature now. The only difference is that nature was then creating species, and trying the best she knew to create as many variations from a given form as it was possible for her to produce. All she can now do is to reproduce the surviving species which she created long ago.

CONCEPTION.—Just what the process is which nature employs in the reproduction of animal life can not be known except by the same mode of reasoning as that employed in determining the beginning of original life, since there can be no starting-point after the union of the first two atoms, which began the formation of the ovum; but when that occurred cannot be known.

If it is true, as Drs. Dalton and Hollock have informed their readers, that the ova were formed prior to the birth of the mother the difficulty of the problem increases, since it cannot be known whether the ova began to form the first day, the first week, the first month, or later on in gestation; whether they all began their existence at the same or different times, therefore the facts must remain a secret with nature.

Doubtless the ova nuclei, be they few or many, are formed prior to birth, and out of zodiacal gases, assisted by the sun and moon for the following reasons:

1st. The zodiacal brain was formed at that time for which reason, I judge, the ova also were formed at the same time, and partly composed of similar matter and produced in connection with certain brain layers at a certain stage of their development. I further judge that the layers were affected by the sun and moon while the ova were forming. I infer this from the fact that it requires 28 days for the moon to complete her zodiacal revolution, which is the menstrual period. Then, after birth and maturity, the moon rules the disposition of the ova, and since it requires that length of time for the ovum to ripen, it is presumably true that the moon is the prime factor in the matter of controling the perfecting of the ova. The

ovum will not develop only to a certain point without
assistance, which I judge must be rendered by planetary
forces, and these forces are supplied in the spermatazoa
which I also judge to be composed principally of
planetary gases, since it is constantly accumulating
in the male organs. When the sperm is received the
ovum takes on new life, and begins a new series of
evolutions.

2d. The idiotic female will reproduce, though her
brain possesses very little of the planetary matter
which is evidence that a highly developed zodiacal
brain is not necessary to the development of the ova.
Thus it appears that nature can create an ovum to a
point that will produce life, even when the planetary
gases are almost wholly absent from the zodiacal
brain of the mother; but developing it to a point
which will impregnate does not imply that it is as
strong and magnetic as nature can make it, any more
than the germinating of a given seed implies that it
is as perfect as nature can make seeds.

Thus it can be seen that there are different degrees
of vitality in the germs of all life. If an abundance
of planetary matter was necessary to produce
ova all intellectual females would produce them
highly developed, since it is presumably true that
the zodiacal brain during their development con-
sume more of the planetary gases than does that of
the idiotic mother, because they produce a healthier
tronger zodiacal brain in their offspring, and attracts
more planetary matter.

I further judge that after birth the condition of the ovum, when perfectly developed, determines in a measure the power of the brain produced from it, and that the quality and condition of the zodiacal brain of the mother at different periods during its development, determines the perfect or imperfect condition of the ovum formed at the same time. I further judge that a well developed vital ova, to produce its best results, must be quickened by a healthy vital sperm. If the male sperm is deficient in any way the ovum cannot produce its best results, since its full force of attraction is not supplied. •One ovum being more perfectly developed and vitalized than another may account in part for the different degrees of strength and intelligence found in children of the same parents, for there is a vast difference in them. Some parents produce some good, some bad, some strong, and some weakly and weakminded children. Other parents cannot produce children above the average, while some women cannot produce them at all, which is evidence that it is not the general strength of the system which gives the reproductive power.

If the statement made by doctor F. Holleck is true, that the average number of ova possessed by females varies from 15 to 30, that explodes the prevailing idea that one ovum escapes each menstrual period. If only 30 ovums exist then less than three years would be sufficient time to dispose of the entire number at the rate of 13 per year. Just where the error lies I cannot inform the reader. Just how and

at what times the ova are formed may never be
known; but doubtless each one is formed during different periods in gestation, and that each is expelled
from the ovaries in the order of its creation, and employed in producing life in the order and in accordance
with the law of its development, for doubtless each
ovum is developed by a seperate layer of brain matter,
and disposed of by the influence of the moon. The
strength of the organism of the individual ovum cannot be known only by the force of the planets operating
at the time of conception. If the mother conceives under benific planetary influences, it is safe to judge that
the ovum employed at that time was well organized,
healthy and strong; but if she conceives under malific
planetary forces, the reverse may be judged. Ova of
different vital forces are possessed by all mothers.

If ovums are not organized previous to conception,
then they must depend solely on the spermatic fluid and
planetary forces for their organization, and also their
condition after birth, which certainly is not the case.

When the ovum passes from the ovaries to the
matrix it is then ripe, and has progressed as far as
nature can develop it in the female organs without
assistance. If, while yet in the womb, it receives
spermatazoa, then new life is infused into its cells, but
if the male principle is not supplied the ovum passes
out and is dissolved. The ovum is sometimes impregnated in the ovaries. If that was always the case
it would account for there being only 30 ova. If
that was true, then no ovum would be forced from the

ovaries until impregnated ; consequently, all perfect ova would produce life ; but from my own
knowledge I can dispute that point, for I have
known them to escape from both single and married
females.

After conception takes place the organization of
the life forces commence in earnest, and the germ now
turns to the electrified walls of the womb for zodiacal
forces to enable it to resume its development. What
the germ demands the heavens supply through the
agency of the mother's brain and blood. The womb
being connected to the brain by nerve branches, the
embrio can receive the brain forces or stellar fluid
directly from the brain and blood since it requires
both of those agents to gather and supply forces to
any and all parts of the body as well as to the womb,
which is the reason why disturbing the womb in its
work affects the brain and why an abortion sometimes
causes death.

As fast as the earth revolves on its axis, the
mother inhales the zodiacal gases, the blood receives
them, and the brain absorbs them and sends them
over the nerves to the womb, where they are received
by the fœtus; thus cell after cell is formed and life
advances.

It is claimed by physicians that the mother's
blood does not pass through the walls of the womb to
sustain the embrio, but only a clear fluid is then received by it. Which being true, makes my argument
still stronger, since the existing embrio must, like the

protoplastic embrio, manufacture its own blood as it is required. The germ from the beginning receives only zodiacal and planetary gases.

The process of growth in the human body is just the same as that in the egg in the incubator, except, Ist, that all of the gases used in the growth of the fœtus must first pass into the mother's blood and go to the brain before it can reach the embrio; 2d, that the egg of the fowl supplies all of the necessary fluid for the development of the chick; the cicatricula, or the little white speck, is the germ, which contains all the magnets of attraction, the white and yellow parts are the food, absorbed by the germ in its growth. Thus the food is stored in the right quantity and quality to supply each division with just what it needs for the magnets to attract the stellar fluids, necessary to develop the chick without waste or overplus.

The human egg, not being supplied with food in the same manner as that of the fowl, requires a special apparatus for that purpose after generation begins, hence the placenta and its contents. After a time the brain grows strong enough to take on celestial gases in sufficient amount to send out a nervous system. The brain being composed of 12 centers throws out as many bundles of nerves. As they advance in length they build the body and differentiate it into human form.

BRAIN IS THE ROOT.—The brain of all animals is the first part of the body to develop, and it grows

proportionately large; but I think no reason has yet
been given by physiologists for this so-called strange
development. The reason I shall assign is as follows:
The brain is the root which throws out the branches
of the nervous system, which are necessary to construct
the physical organism; for without the brain first
there can be no differentiation; consequently, no vital
organs nor body produced. For the foregoing reasons
the brain must grow large that it may store up force
in sufficient quantities to supply nerve-growth to produce the organs and the various other parts of the
body of the fœtus. After the brain had become sufficiently developed to have the necessary strength to
produce the vital organs the pneumogastric and other
nerves were extended the required length for that
purpose, at which juncture organs were formed. The
heart was formed first, at the lower end of the Cardic
Plexus, and the circulation of the blood continued, it
having been established before the heart had received
the impulse of life; finally other organs were formed,
while the embrio continued to grow. After the spinal
cord had extended the necessary length, the nerve
which it encased protruded, and began the work of
forming the vertebra. I think the general understanding is, according to Moses, that the original
frame was constructed, and holes bored in the joints
for the nerves to creep through on their way to the
extremities, as scientific men tell us, for the purpose
of establishing telegraphic communication of the extremities with the brain. While they answer that

purpose, that was not the design. The nerves were sent out from the brain to construct the body, consequently they produced the bones which formed about them in gristly-like substance, but later on ossified, and finally became solid. Then, when the flesh and marrow were removed, the apperture left in the frame appeared as though they were created only to accommodate the nerves in giving them easy passage to the extremities.

At the expiration of the allotted period for the development of the vital organs to a point which would enable them to perform the necessary functions of life, unassisted by the mother, the new being was released from its parental prison to continue life in a new form. Thus was life reproduced, simply, systematically and scientifically, even though man is "wonderfully and fearfully made." So, I understand, physicians claim that the human embrio is a parisite, and has no connection with the mother, because it will sometimes form in the ovaries, and fall down inside of the abdomen, and stick to the walls, and grow outside of the womb, which being true, plainly proves that nature does the work through the agency of the mother, and thus creates the child. It is only necessary for the germ to get access to the mother nerves in order to make a growth. It is not perfect, however, for a placenta, out of its natural case, is too rude to supply perfect conditions, hence the imperfect growth.

THE MENSES themselves are caused by a waste necessarily produced by an unusual effort of the brain in supplying magnetic forces in preparing the germ receptical for gestation, which, doubtless, is to give the necessary strength to the matrix to do its work well. It is known to be highly electrified at this time. And there is no better reason for its condition. The brain not being able to properly magnetize the matrix may, in some cases, be the cause of barreness and the want of the power to retain the embrio through the entire period of gestation, because it has not the power to cling to the walls of the womb.

CHAPTER XV.

BLOOD CIRCULATION.

The heart, we are told, is a force pump used by nature in driving the blood to all parts of the body, thus producing circulation; but if the criminal law will permit it, and the long-suffering public forgive the offense, I will here record a contrary statement to alleged facts, and subjoin my reasons for disputing the high authority that has handed down from the days of Harvey, such a serious physiological error, for if the truth had long been known, the method of treating diseases now might be very different, and perhaps more successful.

In the first place, the heart of any human being does not possess half the power necessary to force the blood one single time to the extremities of the human body, to say nothing of the power required to keep the blood in motion for even twenty-four hours, without considering a long life-time, it must perform this laborious work without rest or repairs; but if the heart does perform this work there is no evidence to that effect only in the fact that the blood does circulate and the throbbing of the heart and the beating of the pulse are simultaneous; but with this evidence it might be a difficult task for the wisest physiologist in the land to tell whether it is the blood passing through

the heart that causes it to throb, or the throbbing of
the heart that causes the blood to flow, if either is true.
The former, however, is more reasonable than the
latter. There is not a man in all Christendom who
would believe that the heart could produce the circulation of blood after he had demonstrated the force
necessary to be exerted by that organ in performing
the work by using a rubber bulb in every way repre-
senting the human heart, to force a fluid as thick as
the human blood through as many rubber tubes of
the same size and lengths as there are arteries and
veins in the human body. Again, if the heart does
perform this wonderful work, whence does it get its
power of perpetual motion. It must be miraculously
constructed to perform all this work of itself, with no
motive power back of it. I believe there is none explained by physiologists. Then again if it does perform this work independently, what power stops its
operations so suddenly, and thereby produces death
in persons who were supposed to be in good health; also
what causes the heart to be the only weak organ in
some bodies and the only strong organ in other human
bodies, since it produces its own force. These are
interesting questions to have answered if the following
explanations are wanting in truth.

I believe, however, that the physiologists have
agreed that there are microscopic capillaries in the
veins that bear the burden of labor, and thus relieve
the heart of that wonderful force it was supposed to
possess. This explanation, however, only tends to

mystify the subject still more, since it is only an assumption, and is not susceptible of proofs.

It is not a pleasant thing for an intelligent man to say he believes a supposed truth without being able to give a reason for his belief, especially when there is none to be given, though he may sometimes be compelled to do so, but to assert that the circulation of the blood is produced by the action of the heart, is certainly without reason or scientific basis. It is hard enough to believe that the heart possesses muscular strength to perform the simple office of a valve operated by a known force, as the valve of a pump is employed by the engine running it in throwing water; but to say that the heart is automatic and produces the wonderful phenomenon of circulating the blood through the system, which would require an incalculable force, is certainly without reason or demonstration.

From all the examinations the doctors have ever made they never have yet been able to discover any special department of the heart for the generating of force, which is necessary to produce circulation; sometimes the heart becomes so very weak that it can scarcely beat at all, and yet circulation to the extremities goes on. The heart will even stop beating for an unnatural length of time, and again resume its operations, which it could not do if self-operated; therefore we feel safe in saying that the heart does not produce the circulation at all.

The blood is automatic, it being composed of

gases, is constantly kept in motion by taking on and throwing off its component elements. When the brain is in health the blood is kept in a highly charged magnetic condition which causes it to move rapidly through the body; but without a regulator it would flow too fast, or be irregular in its motion.

To obviate this difficulty and to produce a regular circulation of the blood through the entire system, nature provided a check valve in the form of the heart, with its chambers of sufficient size to receive and discharge at necessary intervals enough blood to reduce its motion to a proper circulation to supply the demands of the whole system with magnetism, without loss of energy. This organ being connected to the brains, the great magnetic store-house and distributing office of the body, by nerves, its valves are operated by magnetic involuntary discharges from the brain at regular intervals, forcing the valves to open and close, receive and discharge regularly and continuously, thus producing a regular circulation.

Resuscitation in case of suspended animation is strong evidence in favor of magnetic forces producing circulation.

Artificial respiration forces into the blood magnetism, which finds its way to the brain, which revives the organ of life and causes it to throw off magnetism, that finds its way to the heart and operates the valves admitting the blood into the chambers of the heart, from which it is discharged, thus repeating the operation till circulation is restored.

If life depended on the voluntary action of the heart, all efforts to restore life after the heart had ceased to beat would be without success, because the heart is supposed to be the life motor, and therefore could not be forced to beat after once withdrawing its forces. But, on the contrary, it has been forced to renew its action several minutes after life was supposed to have become extinct.

Some physicians might object to this mode of reasoning, because the heart is the first organ to form in all animal bodies. In reply to this objection I will say that it is quite natural and reasonable that it should be so, since it is necessary to have blood and a regular circulation first, and the heart second, which is always the case before the animal can be properly developed—blood first, the heart second, and other organs falling in line of development according to the order of their importance, the brain being the last, perhaps, developed, because the parent of the child performs before birth all the offices of the brain after birth; so this objection is answered, for the brain has no office to perform till after the child is born.

The pendulum of the clock has nothing to do with operating the machinery; it only determines the speed the wheels shall move. Just so with the human heart, nature uses it as a regulator, and not as a motive power.

It is well known that violent exercise will accelerate the motion of the blood, and also quicken the beating of the pulse, not because the heart acts first, for it does not, it has no innate force, and therefore cannot act until impelled to do so by some existing force, and that force is magnetism, which is stored in the brain by the blood.

The action of the diaphragm, which is instrumental in operating the lungs, is involuntary, because it is produced by magnetism sent from the brain. The action of the lungs which causes breathing is also involuntary, because it is produced by magnetism sent from the brain. The action of the lungs which causes breathing is also involuntary, since being operated by the action of the diaphrapm. In the act of breathing the blood is kept in motion by absorbing stellar magnetism from the lungs, consequently human existence is a voluntary action of natural forces, but involuntary human action.

The human organism is so constructed that the functional operations will continue, even while man is wholly unconscious of his existence. The blood carries these magnetic forces to the brain, where they are stored and used by nature's voluntary agent, which operates the whole physical organization from that position, even producing thought and intelligence. The blood being the means of transportation, it is continually receiving magnetism at the lungs and discharging it at the brain. The brain in turn must relieve itself of the surplus force by dispatching it over the nerves to the vital organs as well as all other parts of the body.

This act of the brain, or a portion of it, causes

each organ to perform its functions, whether man is asleep or awake, conscious or otherwise; so the magnetic force from the brain to the heart keeps it in constant action, opening and closing its valves, thus permitting the blood to pass through, by which process life continues. This reduces the voluntary action of the body down to one force, and that is magnetic, the motive power of the universe. This being true, the heart beats fast or slow; first, according to the magnetic condition of the nerve controlling that organ; second, according to the magnetic strength of the blood; and third, according to the degree of mental excitement brought to bear on the brain at any given time. The difference in the number of pulsations produced in any two given persons in health and repose during a stated time, is not very great, which goes to show that nature's voluntary work is not violent. So in order to produce a violent action of the body, the whole organism must become excited, but whether voluntary or otherwise, it must first occur at the brain through the human senses, or by means of stimulants, and not at the heart; which is evidence that the heart is not the prime mover in producing, sustaining life, and keeping up the circulation of the blood.

The heart being connected to the brain by nerves, its valves are operated by magnetism sent from the brain. The blood being highly charged with magnetism gives it vivacity, which causes it to press the valves of the heart, which open, admit, close, and discharge, at the dictates of the brain, when a strong, rapid pulse is felt in a healthy person; but any excitement of the brain will cause it to throw off more force, which quickens the action of the heart, and lets more blood pass through, thereby exciting the pulsations.

The act of running makes the heart beat faster than usual; but the heart is not alone excited by the exercise of the body, all the vital organs are equally stimulated, but the others being not so sensitive to the effect of the magnetic force as the heart is, they are not felt to act with the same power.

In order to produce the bodily exertion, the mind must first be excited, which increases the action of the brain and causes it to attract more magnetism, thus making increased demand on the blood for more power. The brain now dispatches rapidly to the heart for reinforcements, the heart responds to the call by operating its valves with greater rapidity than before, thus permitting the blood to flow faster to the lungs, where it is rapidly vitalized by stellar force, and, receiving new impetus, drives forward, supplying the brain and nerve center with the necessary vitality to keep them in active operation. The brain stimulates the action of the diaphragm at the same time, thus increasing respiration, so that the whole machinery of the body acts simultaneously under the command of the brain. In proportion to the excitement, or the demand for increased forces, will the heart become animated and the whole physical structure become excited and ready for action. The man who is interested in his own business can do much more than his employe with less fatigue, both being of equal strength, because the mind of the master, who is deeply interested in his work, is continually excited, thereby causing the brain to stimulate the heart, which accelerates the blood, thus furnishing the system with renewed strength. But in the uninterested man there is nothing in the business to stimulate the action of the heart; the blood is sluggish, and every movement is forced and labored, and the body soon becomes tired.

In delicate persons the heart sometimes fails to produce a pulsation when it throbs; the reason for this omission is that the brain being weak, it fails to supply to the heart a magnetic current strong enough to open the valve of the heart sufficiently to receive the usual quantity of blood. With no resisting force to check the action of the valve, it comes quickly back to place with a jar, thereby causing a fluttering of the pulse instead of a full beat. This double action of the valve taking place before admitting the blood to the chamber of the heart, will cause the blood to check its flow. A slight disturbance of mental forces excites the brain, which causes it to discharge additional forces, so that when the action of the valve is restored the pulses will become stronger and faster for a few beats, in order to let the surplus blood pass through to the lungs.

Extreme fright will also cause the heart to omit one or more discharges of blood and then beat fast for a while to relieve the pressure at that organ. A fright.

causes the brain to become excited
heart rapidly, causing it to perform j
The most intense fright is caused w'
in a state of perfect composure.
very regular and measured in its b'
slow in motion.

To excite the brain at such a
discharge to the heart magnetic '
cession and causes an imperf(
which prevents the blood from
thus retards its circulation.
caused by this check in th'
many frightened ones to ex(
my throat."

There are two reasons
action of the brain and
in a state of rest and co)
just fast enough to suppl·
forces to keep the m·' ·'
action. In o·'   ·onf·
forces th·   ·uer to su
·' blood must ·
which ca·nnot be· d·····t once.
blood to gather for···· from the lungs, so when the
fright comes the brai··' throws off all of its forces
at once, and then flutters in its excited condition ·
trying to obtain from the blood the gases ·
perform its work. The absor··
the brain ··· · ·

and act on the
ts work in haste.
nen the system is
The heart is then
eats, and the blood

time will cause it to
currents in rapid suc-
ct action of the valve,
entering the heart, and
a peculiar sensation
a circulation has caused
claim, "oh! my heart is in

for the sudden and peculiar
heart, 1st, when the body is
nposure, the blood is flowing
' the brain with the necessary
Aery of the body in a normal
pply to the brain with more
be accelerated in its motion,
it once. It takes time for the
from the lungs, so when the
throws off all of its forces
in its excited condition ·

can utilize. If the excitement would come on gradually the brain and blood would act in concert without producing unpleasant sensations. The blood in its haste to return to the lungs finds its progress impeded by the inability of the heart to pass it through. The heart cannot increase the size of its capacity, therefore, must perform extra labor by increased activity. While this is going on a fulness about the that organ is felt, but this feeling gradually passes away as the brain returns to its natural action, the blood to its normal condition. The body is then restored to its natural strength. A person of delicate organic structure may be frightened to death where the shock is sufficient to destroy the action of the brain cells, and thus suspend the brain force, and prevent it from reaching the heart, which might easily be done where the brain center is weak, and the action of the heart imperfect. The brain may not lose its functional powers wholly, but life goes out before the blood can supply the necessary forces. Artificial respiration would, in some cases, restore life. But when the brain centers which support the heart are active and strong there is no danger of producing death because nature can rally her forces after the shock in time to save life.

Wise men may say that circulation of the blood is produced by the heart, but that explanation is not satisfactory to any scientific man living, since there can be no solution given for its action. If the heart is automatic and supplies its own force, and, consequently, all the forces of the body, then what force can step in and arrest its action for an unnatural length of time and then allow it to resume its functional operations, which it not infrequently does?

But in the absence of the true cause, this explanation has been allowed to follow in the wake of science to satisfy the longing for a perfect solution of the perplexed problem.

It is an easy matter to assert that it is an innate force of the heart that does the work, but to explain a force where none exists is quite another thing. But the magnetic force that rules and controls the action of the universe, impels planets forward in their courses with unvarying velocity, and sustains heavenly bodies in their mutual relation to each other, and establishes and sustains order throughout the universe. Operating in the blood will produce its circulation.

When the forces of attraction and repulsion existing in the molecules of matter can be explained, the action of the heart will be fully understood, the germ of life discovered, and the law of the universe comprehended.

The doctor feels the pulse of his patient, not to ascertain the strength of the heart, but the strength of the entire system, which is produced by the magnetic condition of the blood, though he may not admit it.

If the heart produced the circulation of the blood, then the physician can tell the condition of the heart only by the beating of the pulse, whereas the pulse is the indicator of the various conditions of the blood and body.

same force to the blood, and give a round, full pulse all the time. But, on the contrary, the heart may be in perfect health and the pulse very feeble. The heart may also be weak in health and the pulse strong and full.

Weakness of the heart is caused by a feeble condition of the brain followed by inaction of the nerve, and, finally, by heart failure.

The body may be strong and vigorous in every other particular and the heart weak, but the inaction of the heart will finally reduce the system in health, because it prevents a perfect circulation of the blood, which is necessary to keep up repairs of the system, by distributing material to the various parts of the body as it is needed.

But it is plain to be seen that the action of the heart is not voluntary, but is controled by stellar forces, which are regular or irregular according to the strength of the brain center which supplies it.

The blood may be highly charged with magnetic forces and the heart inactive because the nerve center whence rises its nerve is not strong.

CHAPTER XVI.

SYSTEMS.

It is a singular fact, if the foregoing reasoning is false, that the different systems comprising the human anatomy, are constructed on the plan of **12**, like the new Jerusalem, which the revelator is said to have seen come down out of heaven and rest on 12 foundations, which were garnished by 12 precious stones. The city measured 12 x 12 cubits square, and was 12 thousand furlongs. It had 12 gates made of 12 pearls. They were guarded by 12 angels, with 12 swords.

I do not refer to the foregoing numbers as having any bearing on the human anatomy, but only as an amusing coincidence to show that perhaps the revelator recognized the order and beauty of the heavens, and perhaps understood something of the influence of the zodiac, and allowed his fancy free rein while banished to that lonely isle. Perhaps he fancied that because the zodiac was separated into 12 divisions, there must be something sacred or mystical in that number, and used for a special purpose by the "divine" builder of the universe, and therefore thought he would erect his future city on the same plan as his air-castle, but before it was finished he died, poor man!

NATURE'S MATHEMATICS.—1st. The human frame

is constructed of 17 x 12 bones, or 204.

2nd. From the top of the head to the end of the spinal column there are 4 x 12, or 48 bones.

3rd. On each side of the body there are 12 ribs.

4th. From the end of the finger to the shoulder there are six joints; from the end of the toe to the body there are six joints, making 12 joints on each side of the body.

5th. In the hands and arms there are 5 times 12 bones or 60.

6th. There are 12 pair of cranial nerves.

7th. The eyes are controled by 12 optical muscles.

8th. There are 6, or half of 12, pairs of great vital systems; 1, the bone structure; 2, the muscular; 3, the nervous; 4, the visceral; 5, the circulatory, and 6, skin or superficial circulation.

The foregoing physiological facts ought to be sufficient evidence to convince the most skeptical of the creative laws of nature, but should they be inadequate to satisfy the studious mind, let the critical reader proceed through the next chapter.

FUNCTIONS OF THE NERVOUS SYSTEM.—To coroborate the statement that the nerves control the growth of flesh, another marvelous coincidence, if it be nothing more, presents itself to the eyes of the investigator. It is the inseparable association of the nervous and venous system, for each and every nerve is attended throughout its length by blood veins. The association of these two systems are not accidents either; such an occurrence could hardly have taken place by accident, with two systems so extremely complicated as the two just mentioned, even if accidents are admissable; therefore, it must be conceeded that it was not a slip of the tongue or the toe of Jehovah that caused the association, but a necessity with nature in the prosecution of her work in creating and sustaining the human structure. The body could not be sustained without a nervous system, neither could it be sustained without a venous system.

The question now arises which holds the precedence over the other. It is a well known fact that a large part of the blood can be drawn from the veins without serious or permanent injury to the body, because the absent blood does not affect the functions of that which remains; consequently, if there is enough left in the veins to receive and convey the necessary amount of stellar fluid to the brain, and the various parts of the body, to sustain the organs in the performance of their functions, life will continue, and eventually the absent blood will be replaced. The nerves are more vital in their office because they belong to, and are a part of, the brain; each branch thereof sympathises with all the other branches of that great system in the performance of all its duties. An injury to the brain affects the whole nervous system, and checks or stops the flow of the brain force, which would also stop the circulation; but the blood could not alone be affected by an injury to the brain. If the pneumogastric nerve is parted, death will soon follow. Severing the phrenic nerve will produce a sudden death, because the diaphragm will cease to vibrate, consequently the lungs will stop.

The foregoing facts are strongly in favor of the nervous being the more important of the two systems in executing the functions of life. Moreover, when any nerve loses any degree of its functional power, the flesh it sustains shrinks. When a large piece of flesh is torn from the body it is never perfectly replaced, for the reason that the nerves which constructed it originally are also torn away; therefore, they must be reproduced before new flesh can be formed, but the brain is unable to restore the nerves in their former perfect state, for like shrubs when pruned, they throw out a greater number of branches than they previously possessed, which necessarily shortened their growth. The increased number crowded the newly formed cells and restored the tissue in an imperfect manner, and left a deep scar to mark the place where the wound was made. If the nerves could grow out in their usual perfect form, the healing process would be just the same as the original growth and each cell would be produced in its proper place, and no scar would be left to mark the place where the wound was made.

Without the leadership of the nerves, blood veins could not form; besides, there would be no incentive to push on to the surface of the body and heal the wound. After the veins were once healed, they would remain so, and the blood could advance no further; and since its natural tendency would be to return to but the wound would heal at once without restoring the full amount of flesh necessary. With the nerves growing out they restore new veins, and keep them healed, and also opened, so that they can supply the necessary elements from the blood to the brain force, and thus produce growth as fast as the nerve advances. It is evident that the growth of all parts of the body is produced by the brain through the agency of the nervous system from the following facts:

1st. The flesh will not form beyond the end of the nerves.

2nd. The newly formed flesh is always tender, which is evidence of the presence of the nerves at the surface, but the nerves never protude through the skin, which shows that they do not advance any faster than the flesh grows.

3th. Flesh cannot shrink till the nerves contract, for while the nerves retain their full length they feed and sustain the weight of flesh already produced. They do not double back nor roll up under the skin, but grow and contract in length with the force of the brain.

4th. Flesh cannot increase till the nerves advance. Where the nerves go there the blood veins follows, but no veins are formed apart from the nerves.

5th. The length of the nerve determines the length of the body. Some brains produce long nerves, while others produce short ones, which is the cause of the difference in stature.

The five foregoing facts are quite convincing in their tendency to prove the function of the nervous system, and plainly prove that they are the conductors of the life fluid constantly flowing from the brain to every point of the body

CHAPTER XVII.

ORGANS.

The following table will show the constituent elements employed in the growth of the vital organs. Since the brain supplies energy to all parts of the body, with brain force or stellar fluid, it must be of the same combination of gases as that originally used in the construction of the entire body in order to sustain it in health and activity. The table will show the compound constituent elements entering into the construction of the vital organs. It shows, at a glance, that the elements so abundantly used in one organ is little required in another. In the liver is found 25.23 potash, 1.03 potash in the lungs, and 9.60 of the same in the splene. The blood itself could not supply the necessary elements to all of the organs in the right proportion without some assistance.

	Liver	Lungs	Heart	Spleen
Soda	14.51	19.05		44.33
Potash,	25.23	1.03		9.60
Lime.	3.61	1.09		7.47
Magnesia	0.20	1.09		0.49
Feri-Oixide	2.74	3.02		7.28
Chlorine	2.58	0.00		.54
Phosphorus	50.18	48.05		27.10
Sulphur	0.92	1.04		2.54
Silicia	0.27	0.00		.17
Ferine Phosphorus	0.00	0.00		0.00
Florine				
Phosphorine				

If the brain was a unit in its operation, in pro-
miscuously gathering and supplying stellar forces in-
discriminately to the body, the whole organism would
be unsystematically arranged throughout its entire
structure, and all parts of it would be just alike in
its chemical make-up, which is not the case. As shown
by the above table, each organ is composed of a different combination of elements from the other; therefore, systematic laws must have produced them in the
beginning, and if they did the assistance of the same
laws are constantly required to sustain them in their
functional duties. If all the vital organs were composed of the same combination of elements they would
be exactly alike in their formation, texture and physical
functions; but they are not, and in that fact lies the
secret of their existence, since their difference is necessary to life; each organ being differently formed from
the other requires a different combination of gases to
sustain. Now the question arises, how does each organ
obtain its necessary supply of the necessary chemical
elements to enable it to perform its special function?
Not alone through the agency of the blood can this be
done, since it is a common carrier laden with all the
necessary material for the reparation and reconstruc-
tion of the entire system; but this is not all which is
necessary to keep up the repairs of the body. There
must be a controling agent to attend to the proper

sulphur, 48.05 phosphorus, and so on through the whole catalogue of elements to be discharged at the various organs of the body? This process, as all will agree, cannot be done by chance. There must be a director or a regulator to see that each bone, muscle, cord, tendon, and every organ in the body, is supplied with the required combination in sufficient quantities to sustain the body in the discharge of its entire functions. Nature has attended to that important matter by delegating to the brain this important office. From each of the zodiacal centers of the brain are sent out forces peculiar to itself, and which sustain the parts of the body which it creates. Along the line of nerves extend blood veins, which contain the necessary elements to unite with the brain fluid to build up the body.

CORPUSCLES.—The blood, or plasm, contains the carriers, called corpuscles. They are of two colors— white and red. The red corpuscles, I judge, convey the elements in the blood, while the white corpuscles convey the stellar elements, which they receive at the lungs, to the brain. The brain is constantly discharging their fluids over the nervous system to all parts of the body. When it reaches a broken down tissue, or an absent cell, it draws from the blood the necessary elements to repair the breach, while the blood flows on. It matters not what kind of a cell is missing, or of what combination of gases it is com-

entire physical organism is sustained, and thus are
the vital organs supplied with the necessary energy to
perform their important functions, and thus is each
division of the body supplied with its constituent elements in the necessary proportions without mistake
or delay; but should any division of the brain fail to
accumulate a sufficient amount of stellar forces to
supply the loss of waste matter sufficiently to keep up
a perfect action of any or all parts of the body, the
vital organs will fail to perfectly perform their work,
when the body will decline in strength, and ill-health
will follow.

Thus it has been shown that the animal organism
was mechanically devised, systematically constructed,
and regularly operated by the chemical divisions of
the brain. It is generally supposed that the vital organs were created to support and lend activity to the
brain, as well as strength to the mind. It wfll, therefore, be a surprise to many to learn that the reverse is
true. It is now well understood that the brain gives
life to the whole organism, and sustains the vital
organs in the discharge of their important duty. The
organs cannot cease to perform their functions except
by a cessation of brain force, but when the brain ceases
to perform its functions, no matter how strong and
healthy the organs may be, they will cease operation
at once. A heavy blow on the head will instantly
suspend animation, and all of the organs will cease to
perform their duties, for the jar causes the brain cells
to discharge their entire supply of vital fluids, and

themselves collapse and remain in that condition; but
if the blow is not severe enough to cause the cells to
lose their functional power, they will after a short
time breathing refill with stellar fluid, and life will
continue as before. Sometimes, however, the brain
grows weak from other causes than blows. Then if
the organs cease to perform their functions, life goes
out of the body. Occasionally the brain forces run
down to so low an ebb of life that the nerves cannot
dispose of the waste matter of the body when disease
sets in. Thus an organ may waste away to that extent
that death will follow.

CHAPTER XVIII.

ORGANIC DEVELOPMENT OF THE HUMAN BRAIN.

As previously explained, the human anatomy is composed of different systems, bones, muscles, nerves, etc., each of which is constructed on a basis of **12.** The zodiacal brain is also constructed of 12 general chemical divisions, which being true, it is not improbable that the planetary brain is constructed on the same basis. However, it has not yet been proven.

I did not reach this conclusion from the number of planets which created them, nor from a study of those bodies, but from a study of the attractions of the zodiacal brain and its divisions. It is a fact, well known to the author, that the zodiacal brain at the moment of birth, becomes a magnet of attraction for planetary gases, and eventually produces the phrenological organs; and since the zodiacal brain does attract zodiacal gases, it is presumably true that each division would possess the same power of attraction as the others, and therefore produce the same number of mental organs, since they are formed in groups. There are 42 mental organs now discovered and no divisior will give 12 for a quotient. Four is as near as can be reached to produce that number; but 4 x 12 equal 48, which is 6 in excess of the correct number. If all of the mental divisions have been discovered, then this theory is false, and some other explanation must be given to account for the number known. The power of attraction existing in the zodiacal brain is known from the fact that all the mental groups found in different brains are differently developed, and the various positions and aspects of the planets at the time of birth show to what degree they will develop, which should not be the case if they were not created in that way. The variations of brain growth and development is explained as follows:—The zodiacal division of the brain will not attract planetary gases from all quarters of the heavens except that of Mercury, but certain divisions of the zodiacal brain will attract Mercurial gases from any position he may occupy when he is free from the Sun's beams; but when passing between the sun and the earth, and also when passing the opposition, at the time of birth, his gases are consumed or dissipated by the Sun's rays to that extent that a child born at that time cannot receive them. Other planets are not illy affected by the Sun's beams; on the contrary, the conjunction of the Sun and Jupiter is very good, while the conjunction of saturn and the sun is very evil; but the evil effect appears to be on the sun, to weaken his effect, or, what is more likely to be true is that when they both hold the same position the brain attracts elements from both of them for the same purpose, and the combination does not work well together. All the planets, except Mercury, must be in certain positions before the zodiacal brain will receive them. Another singular fact, known to all the students of this sublime science, is that no zodiacal division of the brain will attract all combinations of planetary gases, as sextiles, squares, trines and oppositions; but each division will attract a certain combination, and utilize them in developing the brain. Thus the trine aspect of the Sun and Jupiter to each other will supply the combination of gases necessary to develop the organs of acquisitiveness adhesiveness and veneration, but the combination produced by squares and opposition of the same bodies will not be attracted to those divisions, and the organs will remain undevelopêd. The trine of Venus and the moon supplies a combination of gases which are attracted to the frontal brain at birth, and develop the organs of order, while the square aspect of the two bodies will supply a combination of gases which cannot be attraoted to that zodiacal division of the brain, therefore order will not be developed. Any aspect of mars will produce a certain degree of development of the organs of time, calculation and constructiveness; but the sextile and trine aspects are much better than the square or opposition. The conjunction sextile and trine of venus and Mercury will produce a combination of gases which will be attracted to given divisions and develop the organs of time, color and ideality; mirthfulness and philoprogentiveness.

The foregoing being true, it cannot be truthfully said that the organization of the brain is the result of a design; neither can it be said that it is the result of an accident; nor yet, the result of heredity; nor parenta\

influences. It is purely the result of natural laws,. which were, so far as can be known, created without design.

Doubtless the laws of nature cause every event of human life; but that they were foreordained by her from the beginning cannot be proven by reason, law or precedence.

HOROSCOPE OF ABRAHAM LINCOLN.

Our martyred and lamented President and Charles Darwin were born in the same year, in the same month, and on the same day thereof, but not at the same hour.

If they had been born in the same latitude and longitude, their births would have occurred two hours apart, which would have left the planetary aspects in their respective horoscopes very nearly the same. The Sun and Mars, shedding their benign influence on each other, gave to Mr. Lincoln physical strength and power of endurance. The Sun, Moon and Jupiter, receiving the friendly influence of each other, gave him hope; Mars, squaring the Moon, gave courage and confidence; Saturn gave tenacity of purpose, while Uranus and Mercury in trine aspect to each other imparted to this great man intuitive knowledge and a clearness of vision, that bordered on to the realm of prophecy. Being born of the same race of people the brains of their parents possessed similar combinations of elements;therefore the planets would affect them similarly if not the same through gestation, consequently, the heavens would create similar results in their respective embrio through the brains of two mothers that they would through the brain of one mother; therefore with a degree of truth it can be said that they were twins. Notwithstanding the fact their fathers were different men and their mothers were thousands of miles apart when these embriotic giants were born; nevertheless they caught the same planetary inspiration. Could Mr. Lincoln have had a twin brother born under the same influence as was Mr. Darwin their horoscopes would have read the same; or, if Lincoln and Darwin had both been born on the "old Kentucky shore," they would have been as near alike as if they had been born of the same mother. They came that near being twins: they were born under conjunction signs, which gave to each different rulers from the other. Mr. Darwin was ruled by Mars and Uranus, which made him wayward, rebellious, despising authority, somewhat selfish and malicious, which caused his mind to wander away from the beaten track of public opinion, and to pass beyond the lines of scientific limits, far into the waste of the unexplored.

Lincoln was ruled by the great Jove or Jehova, with Saturn near his eastern horizon, which gave him a different bent of mind from that of Mr. Darwin. He was the man of the people; he liked social and political institutions; he liked order and social harmony, and though possessing the same great ability as Mr. Darwin, his love of pursuits was different; he chose law for his profession because it was genial to his mind; he was humane and unselfish and placed justice above price. To defend the weak was a pleasure; to advance, and to guard the interests of all was to him a delight. He loved freedom for himself and desired that all should enjoy what he so highly prized. Uranus and Mercury in trine aspect to each other made him thoughtful, profound and farseeing in his plans. His sense of right was less erring than the letter of the law; his justice more discreet than found on the statute books of nations. He could look beyond the act to the motive of the offender of the law and mete out justice according to his deserts; he was a friend to the downtrodden, and his ears were always open to the pleadings of the poor; he knew the right and had the courage to pursue it; he was one of nature's noblest of noble men. In his horoscope his house of friends was badly afflicted, consequently he was left with planetary influences to scale the height of fame. His ability alone won the victory and landed him on the topmost pinicle of fame's dizzy height, there to glitter in the sun of admiration.

CHAPTER XIX.

ZODIACAL BRAIN.

The Encepelon is a peculiarly constructed engine, possessing more power than any engine of like size yet produced by the ingenuity of man. It is so different from any human invention yet produced by him that man has not been able to guess the intricacies of its operation, and not being permitted to see it work in the fullness of its strength, and in the execution of its skillful functions, he can learn but little concerning the force pervading the human brain.

HEADS.—Human heads are interesting objects to study, since the shape of them is determined by the form and strength of the brain.

The heads of all animals also afford an interesting study, since the form of their bodies are determined by the shapes of their heads. All animals, belonging to the same species, are practically the same in form. They are so near alike that the naturalist can classify them from their general appearance with little or no trouble.

The skulls of horses, dogs, beast and birds, as well as all species of animal life, can readily be pointed out by the expert naturalist; yet there is a slight variation in the skull of individual animals belonging to the same species, but which difference is noticeable only to the close observer. Perhaps there is a greater varia-

tion in the form of the human than is to be found
in the skulls of any other species of animal life,
which is attributable to the many and various forces
working in the human brain, and which take no part
in forming the brain of other animals. The skull must
conform to the size and shape of the brain, and since
there are more subdivisions in the brain of man than
in the brains of any other animals it is subjected to a
greater number of variations. The specific difference
found in the human skull is not the result of accident,
nor, as some suppose, caused by an irregular pressure
on a conglomerate mass of brain matter, which, like
so much putty, may be moulded in the various shapes,
as the pressure might determine; but they are all
caused by a natural development of the brain, which
is produced by planetary forces. Some children have
peculiarly formed heads, and their fond mothers ex-
cuse them to those who notice their unusual shapes by
explaining that they were produced by a fall, or lying
too much one on side when asleep, or by some neglect
or carelessness on her part, for the mother will bear
any blame to shield her unfortunate child. In some
cases the ears are thrust well out from the head, and
their tops thrown forward, thus giving them an unpleasant prominence. The mother explains that it
was caused by wearing the cap too low down on the
ears, thus crowding them out and forward; but she
will probably notice that the child thus deformed is
very active, restless, and often cruel and unkind to
animals; also inclined to be rough and wild. He is

not thus endowed because his ears are peculiarly set on the head, but because the shape of the brain which made them stand out so prominently also caused the temperament. The prominence of the ears is the result of natural and not artificial causes.

Ears lying close to the head show a want of energy, whether the child wore his cap low down on them or went bareheaded. Large, well developed or small, well developed heads, and malformations, are the result of natural laws, and all have been produced with each generation from time immemorial, and it is safe to say that the same results will follow during the coming generations. However, it is a hope, devoutly to be wished, that with the increase of scientific knowledge, and a better understanding of the natural causes which produce them, that the percentage of malformations will decrease in the same ratio, and that less suffering will result from personal gratification in the future.

There is a law of growth in the human brain which is not easily understood, and therefore hard to explain in regard to the respective brains born at different hours of the same day. There is a perceptable difference in them, both in size and shape. In their physical divisions they are the same. These divisions embrace cerebrum, cerebellum, and medula-oblongatta, with their subdivisions. These parts are so nearly alike in all human brains that it is not difficult for naturalists to select them from any number of brains belonging to the lower animals. The cerebellum is the seat of life and motion. All involuntary actions of the body are produced by the force sent from this brain. But this is not the point I wish to call the attention of the reader to at this time. It is the general form of the different skulls belonging to the same species, and especially to the human family. This difference is noticeable in the form of skulls born at different hours of the same day. They are produced prior to birth, and usually appear at the rising of certain zodiacal signs. The difference in the developments of the respective skulls is owing to the prevailing elements received from the division which rises at the moment of birth.

ARIES, rising at the birth, will produce a well developed zodiacal, somewhat elongated, but well balanced brain.

TAURUS, when rising at birth, produces a well developed, zodiacal brain, which is broad in front, but not very high, and somewhat inclined to roundness.

GEMINI, rising at birth, generally produces a high brain, narrow at the base, but wide at the top.

CANCER produces a medium-sized brain, inclined to roundness, and generally narrow in front.

LEO produces a large, well developed, zodiacal brain, spherical in form, broader in front than back of the ears, but heavy neck.

VIRGO, when rising at birth, produces a long, narrow brain, arching from front to rear, and a low forehead.

LIBRA produces a small, fine brain, round and rather high.

SAGITTARIUS produces a high, round brain, small at the base, back of the ears, but widens as it rises.

CAPRICORNUS produces a medium-sized brain, more square than round, backhead medium in size.

AQUARIUS produces a large, well developed, zodiacal brain, but usually narrow just back of the ears, which causes them to lie close to the head.

PISCES produces a large, well developed brain, the ears lying close to the head.

The foregoing are the forms of the zodiacal brains at birth, and before the planetary brain is added to them. Notwithstanding there is a noticeable difference in the general form of the human brains, their functions remain the same, except in degree of power. Each division performs its involuntary duty by supplying the internal organs with force to perform their physical functions, but all brains do not execute their work with the same degree of activity, for the reason that all do not possess equal power to generate brain force, and therefore cannot supply it to the vital organs to the same extent.

The difference which marks the forms of the brains, born at the rising of specific signs, is observable even in the heads of twins born at different hours of the same day. One, born at the rising of Libra, will have a small round head, while one born two hours later, at the rising of Scorpio, will have a large, broad head possessing a much greater degree of vitality than the former one; yet the physical divisions and the vital functions of the respective brains would be much alike, except that one would possess less vitality. Why this is true I am at a loss to explain, except it exists in the power of attraction of the ruling nucleus, or the one first formed; it being the organizer of the eleven subsequently added to it. It may have the power to draw them closer, in a more condensed form, in the latter than in the former case. Should the first nucleus originate in Taurus, Scorpio, or Sagittarius, it would be more vital than though it originated in Gemini, Libra or Cancer, for the reason above stated; but why one should be stronger than another degree of the zodiacal belt is not yet understood, but there is a natural cause, whatever it may be, for the difference.

The development of the zodiacal brain I attribute to the predominating influence which the sign ruling at conception had over the germ from that time on till birth. The cause of the predominating influence I further attribute to the fact that the germ contains pervading elements, which would require contact with certain other elements in order to create fecundation, for the simple contact of the ovum and the spermatazoa of themselves would not cause impregnation. If contact was all that was necessary, then any ovum, when ripe, would receive the sperm at any time when supplied, and produce life, which is not the case. Sometimes the germs are so inharmoniously constructed in their chemical make-up, that zodiacal forces cannot induce them to form a union. Should Libra be rising at the time of birth, I should attribute the cause to the predominating influence which that division exer-

cised over the fœtus during gestation. The cause of
the predominance of the Libra elements I would
further attribute to the chemical condition of the
ovum at the time of conception. If it contained a
predominance of the Venus elements it would require
the assistance of Zodiacal elements of the Venus nature,
to unite the two forces, male and female. The division
Libra, possessing such elements to a greater degree
than other divisions of the zodiac, supplies them to
the germ as soon as the earth has reached a point
which make it possible for them to receive the zodiacal
forces required, which would be at a certain degree of
Libra.

Thus it appears that if conception takes place at
the rising of Libra, the birth will also occur at the
rising of the same division, which position would give
the Libra elements the predominating influence over
the ruling nuclei of the ovum, which it would continue
to hold through the period of gestation. In order to
continue this authority over the physical development
of the child, the Libra forces caused it to be born at
the rising of that sign, for the reason that the elements
from the zodiacal division were so strong in the child
before birth that they must continue in power after
birth. In order to do so, the child must be born at
the rising of Libra. The reason for Libra rising is that
the eastern division is the strongest point for the
dominating division to hold at the time of birth.
If by "accident," or by the design of the attending
physician, the birth is forced out of its perfect time,

nature's design is thwarted, and evil results will follow to a greater or less degree; but to what extent can
only be known by the influence of the afflicting planet
in the horoscope of the mother at that particular time.
In some cases the time of birth is apparently postponed, but nature is only waiting for the rising of the
ruling division before producing the birth. The delay
has caused many a mother unnecessary pangs, while
waiting in doubt concerning the result of her confine-
ment, which could easily have been dispelled by
having a little scientific information on the subject
at the proper time regarding the cause of the delay.

Should the Mars elements predominate in the
parental germs at conception, which might be the
case, even with parents who are ruled by the planet
Venus, because the organism of all are undergoing a
constant change caused by the planets moving from
one to another position, from the influence of a malific
to that of a benefic, or vice versa, than it would require
Mars elements to cause fecundation, consequently
Mars elements would be the first to get possession of
the embrio, and when in possession they would hold
the controling position over the other elements, and
rule their development through gestation. Then, according to the law of development, the Mars elements
should rule through life, and, therefore, nature would
cause the birth to take place when the Mars influences
were strongest, which would be at the rising of Aries
or Scorpio.

Parents, born, one at the rising of Aries, the other

at the rising of Taurus, might beget Libra children. In that event they would be very unlike their parents in size of body, form of head, as well as in taste and talents. But should the zodiacal forces, which rule during gestation, say that of Libra, be deposed by a stronger force, say Aquarius, a short time previous to birth, the child would possess the Libra form and features for a while after birth, but eventually it would change to the Aquarius form, and be radically different from what it would have been had it been-born at the proper time. Then, instead of being tall and slender, the child, when grown, would be stout and shorter.

It is thought by some mothers, that if the child be born short of the regular period, in casting the horoscope it would be necessary for the astrologer to understand that fact, in order to make the necessary allowance for the difference in time, but such is not the case. The time of birth, be it longer or shorter than the allotted period for gestation, is all that is required. The first breath of life is the important moment to observe, regardless of what it might have been. When the sign rules uninterruptedly from conception till birth, the growth is complete, the birth is natural, and the form will develop to fill the description of the ruling signs, which should be the case with every birth, and will be when the laws, which govern human life, are properly understood and observed in the reproduction of human life.

HANDS.—Each of the celestial signs, when rising

at birth, produces hands peculiar to themselves, and in many cases the rising sign may be known by the form of the hand and the length of the fingers of the subject; but owing to the varying of the combination of the elements which compose the brain, this cannot always be done.

ARIES, when rising at birth, usually produces a well formed hand, medium in length, smooth fingers, though not tapering. Among the laboring classes the hands are not so symmetrical as they are among students and professional men, born at the rising of that sign. The unemployment of the hands is not the cause of them being small and shapely. It is the condition of the brain which produced them. The sturdy laborer has large, rough hands, not because he labors, but because the brain which made him a laborer also gave him the hands with which to do the work.

TAURUS, when rising, produces a plump, fleshy hand, rather short and broad, and usually the fingers are smooth.

GEMINI generally produces a long, slender hand, with tapering fingers, bony and nimble.

CANCER produces a long, broad palm, which extends out between the fingers, thus producing the appearance of webs.

LEO, when rising at birth, gives long hands, well shaped fingers, and a narrow smooth palm.

VIRGO gives a short, rough hand, with knotty fingers.

LIBRA-hands are long, slender, with tapering fingers.

SCORPIO produces large, strong hands, not always rough and bony, but many times they are.

SAGITTARIUS gives a long, bony hand, but a short palm.

CAPRICORNUS gives short, bony hands, usually knotty fingers, but not always.

AQUARIUS produces a well shaped hand, long, smooth and tapering, medium sized palm, and sometimes it is broad.

PISCES gives a short, plump, fleshy hand and fingers.

The following chapter will be devoted to explaining the growth of planetary brain, or the mind producing grey matter, which is formed after the child is born.

1st. In so doing I must necessarily show that the human mind as an entity, did not exist prior to birth, nor prior to the additional growth of brain matter, which occurred after birth.

2nd. I shall conclusively show that the brain is produced from the gases received from the celestial bodies.

3rd. That the gases received from the planets produce the grey matter, intelligence, and adds to or detracts strength from the brain forces.

4th. That the strength of the mind depends (a) on the size, (b) on the form, and (c) on the activity of the brain.

5th. The small brain, however good, is not forcible.

6th. A badly formed brain, however large, is not forcible.

7th. A large, well developed, inactive brain is not forcible.

8th. That all of the foregoing conditions depend on the local and relative positions of the planets at the moment of birth, and their association after birth.

9th. I shall also show that the brain is composed of chemical, as well as physical, divisions of gray matter.

10th. That the mind is the result of a chemical action of the brain, superinduced by electrical currents from the planets.

11th. That the strength of currents, producing this effect, depends on the position of the planet at any given time.

12th. That when the chemical action ceases, mind will then be no more.

CHAPTER XX.

PLANETARY BRAIN.

The growth of the planetary brain, its peculiar formation, its divisions and subtle functions, embrace, perhaps, the most perplexing part of the entire subject now under discussion, and the one most difficult to make clear to all, owing to the extreme intrincies involved, and the unfamiliarity of the general reader with the subject. But I will try to make it easy, by abridging as much as possible, without obscuring the ideas set forth, and yet indite enough to make clear the theories advanced. The subject, however, is too extensive to be fully embraced and clearly comprehended in a work of this size. The details are too complicated to be fully brought out until the leading points become established as scientific facts. Eventually the metaphysician will find the study of the brain from this standpoint, not only interesting and facsinating, but instructive in all of its details, since he will find a reason for all things human. The brain has never been properly studied for the reason that it could not be reached by physical analysis during life. In death, its functions having ceased, the cells collapse, which places it in a very unfit condition to be studied, to say nothing about the mutilations caused by the knife in dissecting it; therefore, the brain can only be successfully studied in connection with the laws which produced it, which is the author's method of analysis and synthesis.

In discussing the planetary brain, its physical function, being the most important, must receive first attention. The Zodiacal brain having been previously discussed, it is now understood that the planetary forces have no power, independent of the Zodiacal brain, to create life, nor to construct brain matter, therefore the Zodiacal brain must be formed first, and the planetary brain added to it after the birth takes place. This additional growth of planetary matter was and is necessary to assist the Zodiacal brain in more perfectly performing its physical functions. The addition of gray to the white matter after birth must necessarily change the form of the brain from what it was at birth, for the reason that all of the Zodiacal brain centers of attraction could not be supplied with an equal amount of planetary gases all the time, nor could they all receive an equal supply at any given time, because the planets could not hold positions from which they could supply it; therefore they could not all develop alike. Persons born at different hours of the same day, are supplied with different combinations of gases which are attracted to different centers in respective brains, which is the reason they do not always develop alike. It is impossible for nature to create two brains just alike except they are born of the same race of people, at the same place, at the same time.

The planetary brain is composed of two general and many subdivisions. The two general, or to be more explicit, the outer and inner layers of planetary matter which compose the cerebrum and cerebellum will be explained now in a general way and later on more specifically.

While forming, prior to birth, the zodiacal brain has but little power to utilize planetary matter, but at the moment of birth it suddenly becomes transformed into a very active magnet which is able to attract planetary gases, with which to complete the growth of the encephalon. I attribute this change to the fact that the Zodiacal brain, prior to birth, must remain in a passive condition, since it is only in an evolutionary state, and is being acted on by other forces, through the agency of the mother. It, therefore, possesses no volition of its own; but so soon as it is fully developed it is then born and becomes an independent power, and able to create additional growth by attracting planetary gases and transforming them into layers of gray matter.

It must be remembered, however, that all of the twelve divisions of the Zodiacal brain cannot attract planetary gases from the planets, only when they are in certain positions at the moment of birth. When the local positions of the planets are favorable for a given division of the Zodiacal brain to attract their gases, it grows very large, and the increase of planetary brain is plainly visible in all heads, except those of idiots, for they cannot receive it for the reason that

the Zodiacal brain did not receive enough of the planetary gases during gestation to magnetize it sufficiently
to awaken in it, at the moment of birth, the necessary
force of attraction for planetary gases, to produce a
healthy brain, therefore it remains as it was born. It
might not promote the friendship existing between yourself and the newly-made mother to tell her
that her child is an idiot, nevertheless such is the case,
and so it would remain without the additional growth
of planetary brain. All children are born in that condition — yet, after all, it is a well-known fact that all
newly-born infants arrive without intelligence, but the
mental condition of the child at birth is supposed to
be only in an undeveloped state, and, like new wine,
will improve with age, since the brain and mind are
supposed by some be be separate entities; and susceptible of cultivation — a theory not easily explained, but
clear enough to be densely erroneous at best, for age
alone cannot give mental strength to any brain. If it
could, then all brains would be equally active, since
they all have the same opportunity in that respect, to
gain strength; but it is plain that all of them do not
grow alike; on the contrary, no two of them develop
the same. Every degree of intelligence, from the philosopher down to the idiot, is shown in the human
family, and sometimes that difference is noticeable in
the same family. No two brains are ever found just
alike at birth, or ten years later, consequently there
must be some other reason than age or cultivation for
their difference in developments.

The eyes of the newly-born infant are expression-
less; its attention cannot be gained in any way; all it
knows is to kick and cry, both of which a full-grown
idiot can do, but a colt, calf or a pig can do more, for
they can balance their bodies on their feet, run about,
take food without assistance almost immediately after
birth.   If a child should remain in the mental condition in which it is found at birth, no one would dispute its idiocy.   No one disputes the growth of the
brain.   He only questions its conditions at the moment of birth, for it is supposed by all to be as perfectly organized before, as it is any time after birth,
and all that is needed to perfect its development is
time and food.   Generally it is believed that the child
receives its nourishment from the food taken into the
stomach.   Food, of course, is necessary to produce
growth, but most children take food equally nourish-
ing, but no two develop the same intellectually, therefore a more potent reason must be given to explain the
puzzling facts.   If the divisions which produce the
mental faculties are wholly absent at the moment of
birth, which they are, then what force in separate
brains would cause them to take form, and thus produce different talents and traits of character, if not the
planets?   They could not have inherited their peculiarities since they were received after birth.   They do
not will the development they assume, since they have
no power to choose, therefore no power to govern the
form they will take.   It is sufficiently evident for com-
ment that there is a code of laws which rules the

growth of the body, the form of the brain, gives power to mind; produces the events of life, and determines the destiny of the man. All clear-headed reasoners realize these facts, though they may not understand the laws which produce them; but all have witnessed the results of the unseen forces of nature, and know that they exist. It is this code of natural laws which I am now going to explain.

BRAIN LAYERS.—The two layers of planetary brain, now under discussion, are produced by the same planets, but under different circumstances. Both layers affect the strength of the mind and body, but in different ways. Neither layer, under any circumstance, can be perfectly formed, but under very favorable circumstances both may be well formed, or one may be well formed and the other very imperfectly formed, all depending on the local and relative position of the planets at the moment of birth. The first layer spoken of is produced by the local position of the planets relative to a given point on the Earth. The second layer is produced by the aspects of the planets to each other, from the different positions of the heavens at the moment of birth. The latter is the outer layer, and terminates in protuberances called phrenological organs. The Zodiacal brain contairs twelve chemical divisions or centers. Each division has a distinct and important function to perform in developing the intellectual brain after birth. The frontal division attracts gases from the eastern quarter of the heavens. If the birth occurs in the absence of all of the planets from the eastern horizon, the frontal brain will be small. One planet rising increases the size of the frontal brain above the average by depositing one layer of gray matter. Two planets rising at birth gives a greater increase in the size and working power of the frontal brain by forming the second layer; but when the birth occurs at the rising of three planets, they produce immense convolution of gray matter in front of the ears. Three is the greatest number of planets I have ever known to be rising at the birth of any person. [See horoscopes of Grant, Garfield and Butler.] The upper and rear portion of the brain attract gases from the midheavens and develop the sub-layers of that portion of the brain. The lower and rear part of the Zodiacal brain attracts planetary gases from the western division of the horoscope, and develops the sub-layer of the domestic region.

The foregoing explains how the sub-brains are formed.

Aspects.—On top of this sub-brain there are other layers formed by a combination of gases, produced by the many aspects of the various planets. For acuteness of mind and activity of body, the layers thus produced are more important than the sub-brain, for without some of these combinations the body would not be supple. The Zodiacal brain supplies the nervous system with sufficient force to sustain life; the sub-brain adds strength to it, and gives more strength to the muscles, while the organic development further lends strength and activity to the entire muscular system, and gives control over the movement of the body; therefore, the activity of the body is known by the aspects of the planets at the time of birth. To explain the power of the brain over the muscular action of the body, I will present the following illustration:

Beginning with the lowest form of human life, or state of absolute idiocy, or a life produced alone by Zodiacal gases, no physical strength is manifested, even at the age of twenty-one years, for the adult idiot is a helpless mass of human flesh. He experiences no events of life; he has no business ability, no matrimonial inclinations, no vices, no virtues. Every day to him is practically the same; but taking a higher form of idiocy, when only a small amount of planetary matter is added to the Zodiacal brain. The child can stand and walk and is not altogether helpless. He has some control of his muscular action. Then take a still higher form of the same malady. The boy can run, play, and do some kinds of rough work, which requires no dexterity of hand; but not in the least degree is he mechanical. He only has a clumsy use of his body and muscles. But as the planetary brain increases in the healthy child its physical strength, activity, and mental power also, increase; but if there is any difference between them the physical advances beyond the mental powers. When the brain grows large enough to produce organic developments, intelligence is there manifested in gracefully directing the movement of the body in motion. As the brain increases

in size, the body increases in activity. Some men can
walk a tight-rope, lay lengthwise on it; again rise to
their feet, and even push a wheel barrow along in front
of them, and feel perfectly at ease on a single rope suspended in midair, at a distance from the ground, that
some men could not reach on a long ladder without
fear of falling.   Another can ride a bareback horse,
turn a somersault, jump through a hoop, alight in the
proper place, at the proper time, to avoid accident.
The juggler too, will keep three or more knives in the
air, whirling around and around, catching them by
their points, tossing them again, without making a
mistake, losing a knife, or wounding his fingers, which,
however, is not the work of an idiot, but of one who
has time, weight and calculation well developed, which
gives prefect control of the muscular action, so that he
could always throw his knives with the same force,
watch their progress, and be ready to catch them when
they returned.   If the knives were not thrown just so,
they would not return to the right point, at the right
time, to be caught when the operator is ready to receive them; but by being thrown with the same force,
they make the same revolutions in the same circuit, in
exactly the same time, and thus reach the right point
to meet the nimble fingers of the juggler.   The dancer,
too, can make his feet follow the figures of the dance
in the waltz, quadrille, or the clog, and keep perfect
time with the music, and so may many other unusual
feats be performed by certain individuals who possess
the necessary brain development for that purpose, but

not by the clumsy feet nor the unskilled hand of the
undeveloped brain. The combination of the Sun,
Mars and Mercurial elements are necessary at the time
of birth to produce the proper developments to give
power over a special set of muscles for that purpose.

The sub-divisions, or mental organs at the outer
corner of the eyes, are some of the points necessary to
be developed to give such perfect use of the necessary
muscles. It is plain then that the office of these mental organs is to control certain sets of muscles, by supplying them with brain force, and also to direct the movements of the body when set in motion. When large,
these organs create an abundant supply of brain force
for the purpose, and dispatch it to the muscles when
desired, thus supplying them with power to continue
in action for a given length of time. The muscles can
be compelled to obey the will just so long as the brain
can obtain stellar forces in sufficient quantities, and
supply them with that fluid. After a time the dancer
becomes exhausted, because the brain force is insufficient to sustain the action of the muscles; while the
breath grows short and quick, because of the desperate
effort of the lungs to obtain enough stellar force to
sustain the brain in its laborious work, after supplying
the muscles with the necessary power to do their work.
It will be noticed that the brain becomes wearied first,
and at the same time confused in following the figures
of the dance. It is then the feet begin to fail in the
performance of their task, and refuse to perfectly execute their work, and finally get out of time, blunder

and break down. In order to control the feet through the figures of the dance, they must have intelligent directors to mentally outline every line, curve and circle, through which they must pass. These directors can be no other than the divisions of the brain which supply the electric force that operates the muscles which control the legs and feet of the dancer, the rope-walker, and the hands of the juggler. In proportion to the size and activity of the sub-divisions which supply the fluid to any given set of muscles, will they obey the will of the performer for a longer or shorter time. The smaller the organs the less force they supply to the muscles, consequently the less perfect will they control the feet, in the execution of the dance, and the dimmer will be the intellectual lights they shed; therefore, the less accurate will be the forms outlined by the mental directors. The feet cannot better perform than the mind can direct. No matter how large the organs may be, their power is limited, since they can only receive and supply a certain amount of fluid in a given time, which will last a longer or shorter period according to the way in which it is expended. If exhausted by rapid and violent exercise, the time will be much shorter than though the exercise was moderate; but when the cells fail, the vital fluid becomes exhausted, and the brain can supply no more, the thoughts grow dim, and the muscles fail to perfectly perform their work, when the feet grow clumsy and rest is demanded that the cells may recuperate their lost powers.

If the brain can create lines, curves, and circles, for the feet to follow in the dance, the hands of the juggler to trace in his performance, it can also create lines, curves and circles, without making any physical demonstration whatever, because thought can flow, without connecting any branch of the nervous system with any sub-division of the brain. It was the play of electric lights from the above named organs which gave to man his power to invent, combine numbers, in producing arithmetical forms, higher mathematics and architectural designs.

The foregoing are some of the results wrought in the human brain by planetary forces, and plainly show that the first duty of the brain is to perform a physical office, which is shown in the fact that all animals have the use of their bodies, however weak their minds may be, and it is a noticeable fact that the less control the animal has over the movements of the body, the weaker is its intellect. But the brain rules the actions of the body of all life, regardless of its intelligence.

The second duty of the brain is to create mental functions, if they are not simultaneous in their actions. The action of the brain cannot cease; it must continue active from birth till death, asleep or awake, which condition can only be sustained by continually receiving a fresh supply of vital fluid from the atmosphere by breathing. Even after the cells are full to overflowing, the same sources of supply must continue, which would be impossible unless there were some means by which the accummulated forces could escape when the muscles were at rest; but Nature provided for that necessity by allowing the brain to exhaust the

accummulated physical forces in mental action independently of the physical efforts of the body. The brain
could not, like the physical dynamo, shut off its entire
energies and resume them at pleasure, and go on as
before, for it could not have life again infused into its
cells, hence the necessity of keeping them in action
and also disconnected from the nervous system, except
when desired. The connection between the brain and
nervous system is made by a mental effort of some one
of the twelve Zodiacal divisions. The connection is
made for the purpose of conveying the brain force to
the set of muscles to be operated. The medulla oblongatta being composed of a multiplicity of nerves, it
is evident that it has a very complicated and important office to perform. This office I shall compare to
the central telegraph office, which connects the entire
system of wires, so that a wire from any part of the
country may be connected to a wire leading to any lo-
cality desired, and a dispatch sent. Then the connection is broken, and, if desired, connection with another
wire is made, and so on till all of the wires in the office are used. So the medulla oblongatta connects the
sub-divisions of the brain to the desired branch of the
nervous system. In this way any muscles of the body
may be reached by any sub-division of the brain, and
brought in to play by a force of the will or an effort of
that sub-division. Such, I believe to be, the office of the
medulla oblongatta and the method employed by the
brain in controlling the muscular system. After the
dispatch is sent, the force is expended, the connection

is broken, and the muscles released until further used.
The foregoing is the solution I have for the problem,
and the following explanation will reveal my reason
for entertaining it: If the brain was permanently
connected with the entire nervous system, the flow of
brain force over the nerves would be constant, therefore the muscles would be kept continually in motion,
and there would be no rest, day or night; for each sub-
division would alternately send a dispatch over the
nerves, and thus keep some of them active all of the
time. But, as it is, there is no connection without an
effort of the sub-division of the brain, which sends the
dispatch. Any and all branches of the nervous system can be connected to any given sub-division of the
brain at will, which is shown in physical action. A
single sub-division of the brain may be connected
with one-half of the entire nervous system, which is
proven in the act of controlling the body by the organ
of time. The force from that sub-division can keep
the legs, arms, and the entire body in perfect time
with the music. That all parts of the body have full
connection with the brain is further evinced by the
development of the hands, which have been discovered
as indices of the brain, are caused by the development
of the muscles which are produced by the branches of
the nervous system which develop the hands. It is
true that each sub-division supplies its part of the
matter for developing the hand, since the muscles are
thrown up and make certain developments according
to the force sent from the brain, and thus createsg

mounts, lines, and other marks, which are indices of
brain power. It further appears that the muscles develop in layers, and that each sub-division produces
its own layer. If they are well developed and strong,
the layers are many, and the hand is well developed
and highly marked with lines, stars, crosses, and other
symbols of mental strength; but if the mounts are absent, and the palm is badly marked, the hand indicates
a badly organized and weak brain, incapable of great
effort or business success. If there were no connection
between the brain and the hand by the nervous system, there could be no indication in the latter of the
strength of the former. If the nerves do not form the
muscles of the hand, there can be no active connection
between the brain and that member. The
connection of the brain with the pedal extremities is
just as perfect as that of the hands; but the muscular
development of the feet is very different and requires a
special study. The brain of an infant is proportionately large, and supplies a superabundance of brain
force, which causes the mind of the healthy child to be
very active, and also keeps its muscles active; therefore, the body is in a constant state of unrest. This is
why educators say that activity is a law of childhood.
As the brain increases in strength, it is thereby enabled
to throw out its force, feed the nerves, extend them a
little farther on, and thus produce growth each day.
When the nerves have reached the full length, the
brain has power to extend them, the mind becomes
less active, because the full force of the brain is re-

quired to sustain the body it has created, and restore the broken down tissue that is continually wasting away. Thus the planetary brain assists the Zodiacal brain, in producing growth and muscular development.

The Zodiacal brain, having a permanent connection with the nervous system, it is constantly supplying a mild flow of vital fluids to the entire body in sufficient quantities to restore to the muscular cells their exhausted vital energies, and thus keep up the repairs of the body, without producing any visible effect on the muscular system while it is in a state of rest. That the Zodiacal brain does produce this office is shown in the case of idiots, their muscles being very inactive for the want of planetary brain to supply the necessary current of force to give them strength and activity. The Zodiacal brain also supplies the necessary forces to the vital organs to sustain them in their labors, without any assistance of the gray matter. Should the Zodiacal brain become detached from the nervous system death would instantly follow, which is the reason that the central portion of the brain is more vital than the outer layers. Any disturbance of the medulla oblongata will cause instant death, because the breaking down of its fibrous work shuts off the supply of brain force from the vital organs, when the whole machinery of the body stops.

There are two hemisheres of the brain, each possessing the same number of like sub-divisions, which perform like physical functions, so that each side of the body may be controlled by separate divisions of the brain, which make both sides work in harmony together. The nerves, sent out from the right side of the brain, cross over to the left side of the body, and each half of the brain creates and controls the opposite half of the body, which gives balance and poise to the entire body. By this arrangement of the brain, one or both sides of the body may be put in motion at the same time, without interfering with each other. Where the organs of time, which control the regularity of the movements of the body, are developed the same on both sides of the forehead, then each supplies force to its side of the body. If these organs are well developed, then it will control the actions of the muscles in any way desired. Thus it can be seen that the body is double, and is composed of two distinct organisms, which is further known from the fact that each side of the brain produces its side of the body, which gives pairs instead of single organs and members.

On each side there is one eye, one ear, one nostril, one leg, and one arm. The mouth is also double, it having the opposite half of each jaw like the other; the teeth are formed in pairs; the tongue, split lengthwise in the centre, would produce two halves exactly alike. The dual organism of the body is further evinced in the internal organs, for there are two lungs, livers, kidneys, and a double heart; other internal developments might also be classified in pairs. If both arms depended on a single sub-division of the brain for its motive power, only one could work at a time, and the other must remain at rest; but, since each is ruled by a separate sub-division, both can work at the same time without interfering with the other; for each also has its own mental director to control its every movement. To all outward appearances the phrenological organs and the following named physical divisions, cerebrum, cerebellum, and the medulla oblongata, number all of the mechanical divisions of the brain, or those that have special functions to perform.

Had not Dr. Mayer, an eminent chemist quoted by Dr. Flint in his late work on physiology, by a very careful analysis discovered seven layers of brain matter, it might be a difficult task to convince the general reader, and perhaps would not be a safe experiment to inform the physiologist, that a greater number of layers yet exist to be discovered in the same way before the doctors are familiar with that marvelously constructed organ. These layers cannot be traced out by the unassisted eye of the most astute anatomist. It must be aided by the finest instruments known to the profession. And, even then, they cannot be discovered by any plodder. These important discoveries, made by Dr. Mayer, were wholly due, he claims, to the peculiar form of the cells which compose each layer, but why they were thus formed the doctor did not deem it necessary to inform his co-workers. In making this discovery Dr. Mayer added another feather to his own plume, and innocently won a brilliant victory for astrology and spontaneous production.

CHEMICAL DIVISIONS.—I will cite the reader to the

more subtile divisions of the brain, and discuss their
forces and functions as they are manifested in life. At
one time this would have been a very unpleasant task
to perform, since there was no evidence to support any
statement which might be made concerning the existing layers of brain matter; but since the wonderful
discovery by analysis has been made by Dr. Mayer, I
can now proceed with more confidence in my effort to
establish the fact that many more than seven layers
exist, and moreover, I shall do what the learned doctor, however wise and expert he might be in his profession, could not do, and that is to explain how the
chemical layers were produced, and also explain their
functions which have never been understood.

BRAIN LAYERS.—If their were no planets to affect
the Zodiacal brain, then each division of the heavens
would produce the same effect on all brains born at
the rising of any given division of the Zodiacal belt,
but since there is a marked difference in all persons
born at the rising of the Zodiacal division, as the following will explain:

The motion of the earth on its axis is necessary, in
order that the planetary gases may be supplied to the
Zodiacal brain after birth, so that the sub-divisions
may be produced, and also the layers thereof; for the
sub-divisions are composed of layers of brain matter.
The first day after birth the nucleus of the first layer
of each sub-division is formed. The second day after
birth the second nucleus of the second layer in each
sub-division is formed. The third day the third nu-

cleus is formed. And so on till the whole nuclei are formed, which are to create all the layers of brain matter, after which growth is added to each nucleus until the brain is perfectly developed. The number of layers of brain matter thus formed cannot be known; however, there are as many layers as there are years of life; but how many more cannot be told. It is singular, but true, that each layer of brain matter above described has a vital effect upon the working power of the Zodiacal brain which they affect the most, for each sub-division of the brain does not affect all of the Zodiacal divisions. The Zodiacal division which creates any given sub-divisions and their layers are alone affected by them. The strength of the sub-divisions and their layers are known by the positions and aspects of the planets which produce them. These layers produce the events of life, and cause them to occur at stated periods. The brain layers cannot all work at the same time, but each must take its turn in the executive department of life, and dispense the vital forces of the brain for twelve months, after which time its physical functions are never again required for that purpose. Each layer, however, must retain the record of all the events that occur during its reign of twelve months, as business events, sickness, accidents, births, deaths, etc., and be able to reproduce them when called upon to do so, which act of the brain is called memory.

That a single layer of the brain rules the events of a single year of life is known from the fact that a child could not live if the layers must all work at the same

time, because they are not formed at birth, and therefore cannot all be present in a newly born infant's
skull.   In fact, the nuclei are not all formed short of
ninety days after birth, and even then none of them
are perfectly grown.   The organization of the first
layer determine the events of the first year of life.   The
organization of the second layer, determines the events
of the second year.   The third layer the events of the
third year of life, and so on.   The tenth layer the tenth
year, the twentieth layer the twentieth year, on to the
close of life.   The positions of the respective planets
and their aspects to one another each day, after birth,
determines the mental and physical quality of the
layers of matter they produce; for no two of them are
constructed just alike, owing to the planets changing
theis local and relative positions, thus daily and hourly
forming new aspects, and supplying new combinations
of gases for the construction of brain layers.   Each
layer thus produced is strong or weak according to the
chemical combination composing the nuclei.   If the
combination is harmonious, then the nucleus formed
will produce good results, and a year of good events
will be produced by that layer during its reign.   In
order to know the quality and the character of events
they will produce, and the time when they will occur,
it is only necessary to know the positions of the various planets on the day the nucleus was formed, and
the character of the planets which produced it, or them.

CHEMICAL LAYERS.—The Moon moves faster than
the Sun or any of the planets; consequently in passing

around through all of the divisions of the Zodiac, she must pass all of the planets before completing her revolutions around the Earth. When passing the benefic planets, they produce good combinations of gases which the Zodiacal brain attract, and utilizes, for benefic purposes, in forming the nucleus which will later on develop into a layer of gray matter. While the Moon is passing evil planets, their nuclei which will produce evil results will be formed. The gases from the Moon and Jupiter are harmonious, and when those two bodies are joined together, or are forming harmonious aspects, the Zodiacal brain will attract their gases and produce strong, active, and health-producing layers, with business producing forces, journeys, gifts and social reunions.

When she passes Venus, the combinations of gases thus supplied to the Zodiacal brain is utilized in producing a layer of matter which will bring new friends, introductions, social advantages and marriage.

When the Moon passes Saturn, the combinations of gases from those bodies produces a badly organized nucleus, and brain cells that store up malefic forces, which produce unfavorable events in the years corresponding with the number of days in which they were produced, as ill health, accidents, losses, disputes and law suits. Should the Moon pass two planets in a single day, the layers of brain matter so formed will not be mixed, but they will produce a mixture of events during the year corresponding with the day on which it was formed. Should she pass Mars and Jupiter on

the thirtieth day after birth, then, during the thirtieth
year of life, severe sickness, would be followed by success in business, gifts, or legacies.

The above described motions of the Moon are
called directions.   They are of short duration, and
last but a few months.   There are other directions which
last much longer, consequently much more severe.

The apparent motion of the Sun is much slower
than that of the moon, but faster than Mars, Jupiter,
Saturn or Uranus; therefore in the course of time he
must pass all of them.   He requires eight or ten days
to pass either of them.   These days represent the
years when the events will occur.   The direction of
the Sun to Saturn causes a depression of spirits, a low
ebb of life, sometimes ill health, heavy losses, and even
death, may occur, during the evil transits of the malefic planets.   The Sun passing Jupiter brings successes
for eight or ten years, and especially when Jupiter is
making favorable transits during this time.   Any of
these directions may occur, early, in middle, or late
in life, all depending on the positions af the planets at
the time of birth.

Should the evil direction occur at thirty, forty-five,
or sixty years of age, unless the horoscope is very
strong, heavy losses, serious sickness, or death, will
follow, because the brain is only a dynamo which is
kept active by the stellar and planetary forces.   By
t he planets changing their positions in the heavens,
these forces will increase and also diminish in their
effect in this human dynamo.

The Zodiacal brain has power to create only so many layers of planetary matter. When that is done the growth is complete. The farther the layers extend from the central brain, the less vitality they possess, till finally it runs out altogether. No matter how vital they may be near the center of the brain, they grow weaker, as they recede from that point, until they are able to sustain life no longer, when the mental dynamo stops. Even when the full length of the golden thread is reeled off, life cannot extend very far beyond the century mark.

Some brains are more vital than others, owing to the very favorable maternal and planetary conditions which organized them, and will live much longer than those less fortunate, but even their years are numbered.

Thus, by studying the laws of nature, facts are snatched from the realm of mystery and the ignorance and folly of man is made plain; his duty clearly pointed out; his relation and responsibilities to his fellows permanently established, and the justice of nature extolled. It further proves that the "sins of the father" need not be entailed upon the second, much less the third, generation; for, as a child is born, so will it be; "as the tree falls so it shall lie."

CHAPTER XXI.

LAW STILL IN FORCE.

It seems never to have occurred to evolutionists that the laws which produced man in the beginning, are just as necessary to perpetuate his existence now as they were to create him originally, or they would not so persistently dwell on hereditary influences, parental causes and social environments to explain mental and physical effects. How easy it is to believe, that after nature produced man, she withdrew her forces, and left him in his feeble condition to battle with the elements, animal ferocity, and his own, the most unkind of all, and thus make the most of his very unfavorable surroundings and even claim for him the right, and insist on him exercising the power of free agency. But how they have veered from the direct course of facts. Instead of nature withdrawing her forces, she continues to exercise her power with the same unabated energy as before, and still the work of reproduction goes on, the same as the original process of production, only in a slightly different way. Man thinks he is responsible for the existence of his species. He does not realize that he is a creature of law, and has no more to do with perpetuating his kind than he has with producing a tape worm. His reproductive functions, which nature gave him at the beginning of his life, is simply to preserve the original condition which she established at the time the first human pair were incubated, and she does the rest. If she withholds the seed from woman, or the fructifying forces from man, they possess no power to generate life; but if nature imparts the necessary conditions, then children or tape worms may be produced. The only rejoinder to be brought against this argument is that there are more children than tape worms brought to light, and that the worms appear without an effort on the part of the producer.

The animal heat of the mother supplies the germ with the necessary warmth, which the earth originally supplied to the protoplasm, while the male supplies the fructifying forces which the planets originally supplied to the protoplasm to awaken in the ovum the sense of attraction for stellar gases, and thus is life generated. But the magnets thus created, not being in a position to attract directly from the atmosphere the stellar elements, as the egg does, or the protoplasm did, they must necessarily obtain them through the mother's organism, by adhering to the walls of the germ recepticle, and there absorb the necessary elements from the mother's brain and blood. Finally, after weeks and months of growth, the new life was thrust upon the world.

WHY CHILDREN DIFFER FROM EACH OTHER AND THEIR PARENTS.—At conception the ovum contains all the elements belonging to both parents at that time. But the blood or the brain of the parents never possess exactly the same combination of stellar elements which they had at the moment of birth, for the reason that their respective divisions of the brain cannot collect the same combination of gases all the time, owing to the different positions occupied by the ever changing planets at different times; therefore, they cannot supply the original combination to the embrio at, nor after, conception takes place, hence the difference. The laws of nature, and not the development of the parental brain, only to a limited extent, determines just what combination of gases the embrio shall receive at any given time; but, since the planets were holding a different position at the conception of the child from what they were at the time of birth of the parents, it must necessarily differ from them as much as the combination of elements which compose them differ from those which compose it. The quality of the parental brain has something to do with governing the time of conception, consequently with the time of birth, under the following circumstances: The mother who has a large well developed brain when under favorable planetary influences, will not conceive until the planets are in good position to produce good results, hence good children are born to her. Conception at such a time is due to the fact that at some previous time a vital ovum was formed out of a good combination of gases, and therefore would not impregnate until it received the required combination to elements to produce a certain effect in the child. Then, after conception, the mother being under the influence of the benefic planets, they continue to supply their gases while the embrio evolves until it develops into a strong healthy child. Another ovum, formed at a different time out of another combination of gases, would require different influences to produce impregnation. There is no denying a tendency toward heredity, but it is only a tendency and not a law. If it were possible, nature would occasionally develop in the child the counterpart of one of the parents; but she never does. But what is more puzzling to the casual observer is that no two children are just alike. Even twins are different; but if born very near the same moment, they are very much alike, if of the same sex. In that case, they would be very unlike in appearance and the events of life. When the sex is the same their events are generally the same. In three cases have I known twins to die at the same time. 1st, by drowning, 2nd, by the same bullet, which passed through their bodies while standing near together, and the third from disease. I have read of twin sisters marrying twin brothers, and of twin brothers marying at the same time.

The visible cause of children resembling and differing from their parents in form, features, and events of life, is due to the Zodiacal and planetary laws, and explained as follows:

A father, who was born at the rising of the sign Leo, would be large and light complexioned. Should he have a son born at the rising of the same sign, the on would resemble the father in personal appearance,

but not in mental qualities. Such a birth might easily
occur, since the sign Leo rises once a day.

Again, a father might be born at the rising of the
Sun in the sign Leo, which would make him large,
light complexioned, with light curling hair, with large
bones and a strong frame. As the sun rises in Leo,
from July 24 till Aug. 24, a child might be born to
this father during that period of any year. In that
event the child would resemble his father. In so far
as the influence of the Sun could effect the development of the brain, he would be like his father; but if
born at other hours, the child might not resemble the
father at all. A father might be born at the rising of
the planet Jupiter, in the sign Gemini. Jupiter requires twelve months to transit through that sign,
which occurs once in twelve years. Then, if a child
is born to that father, during his 24, 36, 48, or 60 years,
at the rising of that division it would resemble him in
personal appearance, but they might widely differ
otherwise; for, as the position and their aspects of the
planets differ in their respective horoscopes, just in
that particular would their brains differ in their development and mental activity. The influence of the
Moon at the time of birth has a great deal to do with
forming character. If she was very weak, by her position and aspect to other planets, at the birth of the
father, she might also be weak in the horoscope of the
child. Her motion is very rapid therefore, by nature
hastening or retarding the birth a few days to accommodate the hereditary influences so that the child could

be born when the Moon was in a similar condition as
when the sire was born; consequently the child would
be much like its sire in that particular.

Children resemble their parents, because they are
born at the rising of the same Zodiacal sign, and that
some of the planetary influences are the same in their
respective horoscopes. All persons who are born at
the rising of Leo, or any other sign, resemble in personal appearance, and sometimes we find them so
nearly alike that they might be taken for twins instead of strangers. This is the reason we sometimes
see such a marked similarity existing between strangers
and why we can see in every town some one who re-
minds us of friends or acquaintances of former days
in distant places.

A mother might be born at the rising of the sign
Libra and have one or more daughters born at the
rising of the same sign. In that case they would all
markedly resemble each other.

If given parents were born when benefic planets
Jupiter, Venus, or the Moon, were in the midheaven,
they would be very fortunate in business matters connected with the general public; and also in their social
relations, should a child be born to them at a time
when one or more of the benefic planets were in the
midheavens. In its horoscope it would resemble its
parents in the foregoing particular, and, like them,
would be very fortunate in business matters and in
social relations.

Now, while the midheavens in the horoscope of

the parent is strongly fortified by benefic planets, let us suppose that the seventh house of the horoscope which rules their marriage, was badly afflicted by Saturn. In that event they would be very unhappy in their domestic relations; this would be as strong a point for evil as the other is for good.

Now, supposing the mother should come under malefic planetary influences some months prior to the birth of another child, the forces thus received would produce an evil effect in it, and would cause the worst elements in the parents to develop in the child, which is the domestic nature. When the day arrives for the birth to take place, nature has designed that the house of marriage in the horoscope should resemble that of the parents, therefore the hour of birth is postponed until the Earth has reached a point which will place one or more malefic planets in the seventh houses of the child's horoscope; consequently, the domestic organs will remain undeveloped, and the child, in his later years, will be very unhappy in his domestic relations. Again, supposing that while nature is selecting a time, which will make marriage very unfavorable for the child, it is not necessary, according to heredity, that the division of the brain which rules the social and business successes should remain undeveloped; but nature, in selecting the hour to produce the necessary afflictions in the seventh house, it so happened that Mars must be elevated to the tenth house, from which position he causes losses and troubles and a want of enterprise and public spirit; consequently, he would not only be unfortunate in marriage, but in business also. His friends would think he inherited his domestic troubles, but they might have to go back a generation, or so, to explain the cause of his business failure. In another child, when the influences controling the mother were favorable and the inherent forces were trying to create strong public tendencies in the child, it might be, in order to do so, that the benefic planets would fall in the seventh house. In that case the child, contrary to parental forces, would be fortunate in his domestic affairs. Again, supposing a very ordinary couple should come under very powerful benefic planetary influences, which they sometimes do, and during which time a son is born. In such a case nature would select the most favorable hour of the day for the birth of the child, and the result would be a superior brain-development, not necessarily like its father, mother or any of its ancestors, but perhaps superior to any of them, which in some instances is the case.

No matter how brilliant the parents may be, they cannot produce healthy, happy, fortunate children when under the influence of malefic planets, since they have no power to impart their natural talents to their offspring; they can only transmit that which they receive from the planets during gestation, the forces simply pass through the brain, over the nervous system and out through the pores of the skin. Nature can usually select an hour for the birth when the planets can produce a child much like its father or

mother, but never exactly like either of them. Thus
it can be seen that the physical tendencies may be to
reproduce parental talents and traits of character, but
when the heavenly hosts are arrayed against it, blood
will not tell.

Great mental powers do not indicate great reproductive powers. If any difference exists, the reverse is the case. If the quality of the offspring or
the entire absence of children from the homes of noted
men and women is any evidence, there is enough
to prove that assertion. Where are the descendants
of the great warriors, Alexander, Cromwell, Napoleon,
Wellington, Lafayette, Jackson, Grant and Lee.
Orators—Webster, Douglass, Thad Stevens and Henry
Clay. Authors—Shakespeare, Byron, Longfellow,
Bryant, Dickens. Philosophers—Ptolemy, Coperna-
cus, Kepler, Newton. Inventors—Stevenson, Fulton,
Hoe, Whitney, McCormic, Morse, Howe and Good-
year. Reformers—Luther, Pain, Bennet, Darwin,
Kant and Voltaire. Preachers—Even the saintly
mouthpieces of "Jehova" show no better results in re-
generation. What can be said for the offsprings of
Penn, Cotton, Matthew, Peter Cartwright, Surgeon,
and the famous Beecher family. Queen Victoria is a
famous woman but an unvenerable mother. Teny-
son's mantle did not fall on the shoulders of any of
his progeny. Fields left children to disgrace his
name.

Again, where are the ancestors of the famous men
and women; they cannot be traced back to pinicles of

fame. Then, for what is all this talk of ancestorial blood. royal strains or evolution? It is a healthy, lung exercise, but nothing more. It is neither scientific, philosophical nor true. Children may be inferior to their parents, but they are always equal to the laws of their creation, or they may be superior to their parents, but never superior to the laws which created them; they are only superior to the incubator which protected them during their evolution from the ovum to human life. But may the good stars ever shine on that animated machine.

CHAPTER XXII.

NATURAL LAWS.

The forces which rule the universe are familiarly called Natural Laws, and notwithstanding the familiarity of men with them, nevertheless they are strangers conversing with scientists in unknown tongues. These laws embrace all the forces which produce the many aspects observed in nature. When a strange phenomona appears in any one of the three kingdoms, it is said by scientists to be the result of a natural law; which explanation is satisfactory to all. But so far as they have yet been able to enlighten the world they only know the laws by their effects, for they cannot tell what will be the character of the next phenomona produced by the same laws.

It has been said that fools can find effects where philosophers fail to explain causes. Nevertheless, causes exist, and the only way to discover them is to diligently search for them, and not refuse to look because some one laughs. They can laugh, but fools never patent inventions, discover obscure facts nor apply great principles, yet who will deny that the laughing, sneering public does not to a marked degree rule the scientific world? When the time comes that the "common herd" will laugh at all men who do not believe in Astrology, than every man holding a public position of importance, will subscribe for the leading Astrological journal, and proudly tell how often they have had their horoscopes read.

The Poet Lowell wrote:

Truth forever on the scaffold,
Wrong forever on the throne.
But that scaffold weighs the future.

PHYSICAL ENDURANCE OF ATOMS.—Atoms of matter, like their combinations which enter into the structure of the human body, can only work for a limited time in maintaining the physical structure when their vitality becomes exhausted; they are released from their duties, and other atoms take their places at the proper time to sustain the physical organism. The deserting atoms at once become negative forces and doubtless return to the place of their origin to recuperate their lost energy, for matter never dies. The loss of energy may be accounted for by a separation of their component elements under those peculiar conditions, but glasses are not yet strong enough to reveal all of the excentricities of gases, each of which is now supposed to be a single element, but they may be discovered to be composed of many, the union of which gives them life-sustaining forces, and their separation life-destroying powers.

The energy of individual atoms of matter is limited, consequently they must have rest, therefore when they relinquish their hold on the human body, other atoms must be within easy reach to take their places.

Fresh atoms find access to the body by means of

food, water and air. The heavenly bodies supply the
necessary force to sustain the wearing away tissue, but
elements absorbed from the food taken into the stomach
are dead matter until they are vitalized by the stellar
gases, after being carried by the blood to the lungs.
Food elements cannot be utilized by the brain until
they are vitalized at the lungs, when they become life-
sustaining forces and are sent from the brain over the
nervous system to all parts of the body. That the
vitalization of the food elements is necessary, is known
from the fact that strong men will starve to death for
want of oxygen when well supplied with plenty of good,
wholesome food, if they are confined in an airtight
room.   Men will live more days in the open air without food than they can hours confined in a closed
room with the very best the market could supply.

The parts of the human body that will
waste away and are again restored, which changes may
often occur during a lifetime without producing any
local or permanent injury.   In fact, these changes
constantly take place within the human anatomy;
but since the waste is being restored, the effect is
not noticed except at certain times when the brain is
unable to make the necessary repairs.

The parts of the body subject to changes are the
nerves, fats, blood, muscles, etc.   The fat man may
grow lean and the lean man may grow fat, while large
quantities of blood may be taken from the veins without serious injury to the general health of the body,
the nerves will increase and contract in length, the

nails and hair will continue to grow, teeth will wear
out and disappear. There are other parts of the body,
however, who will not change from a natural cause;
when once produced they remain the same till the close
of life, except in case of disease. Some of these parts
are the brain, lungs, liver, heart, kidneys, bones, and
viscera. If, from disease or a wound, the lungs are
partly destroyed, the wound may be healed and health
restored, but the lost tissue cannot be replaced. Part
of the brain may be removed by violence and the
wound healed, but the cells as they originally existed
cannot be reproduced. The internal organs are subjected to the same unvarying law, for they cannot
waste away even by starvation; their tissue remains
the same till death.

The foregoing facts go far to prove that the internal organism is not undergoing the constant change
we have been led to believe that it is, which leads to
the conclusion that the combination of gases, which
nature employs in the construction of the vital organism, possesses a greater degree of vitality than those
which entered into the construction of other parts of
the human anatomy. No doubt, the atoms, orginally
employed, continue in their places until their vital
forces are exhausted, when their functions cease and
death follows.

The combination of gases received from some of
the Zodiacal divisions possess a greater degree of vitality than others do, for which reason life continues
longer in one than in another case. The fact that

the brain-cells, after being destroyed, cannot be restored discloses the sophistry of the phrenologist, who
teaches brain cultivation by constant study. If nature
cannot restore lost tissue when removed by violence
then study will not increase the size of the natural
brain, nor add one single cell to its original growth.
Study does not increase but exhaust brain-force; it
does not destroy cells, but study exhausts the fluid
which they supply. When the brain is strong, the
cells are active and, therefore, kept full. In such cases
study will not affect the organism, but if the brain
cannot obtain from the blood a sufficiency of the vital
fluids to supply the necessary force for the operation
of the entire machinery of the body; then study will
exhaust an undue proportion of the vital forces, and,
in consequence thereof, the entire physical system will
suffer.

The cell forces at work in two given brains
may be compared to two streams of water flowing
through two pipes of the same size. One pipe is forced
by a very small, and the other by a very great head of
water. The first is sufficient to keep the pipe full with
a constant stream flowing through it but not enough
to give any great amount of force, while the latter
sends a stream of water through with force enough to
run a factory. Both pipes are equal in size, but the
difference is solely due to the weight of water at the
head. So it is with brains of equal size and possessing
equal cells; they will produce different results for the
reason that one has the power to keep its cells full to

overflowing, while the other has only force enough to sustain a feeble current of life; hence the difference in their mental and physical powers. The stronger brain can work overhours with little fatigue and a short rest for recuperation, while the weaker brain is always weak and sluggish. It has frequently been argued that the brain can be developed by constant study, because the muscles are increased in size and strength by labor; but perhaps the philosopher, who advanced that notion, was not aware of the fact, that the muscles are supplied with vital fluid and physical forces from the brain cells, and that every ounce of increase in their size and every pound of strength added to them is obtained at the expense of the brain, therefore it exhaust in proportion to the amount of matter supplied to the building up of the tissue of the muscle.

Muscle cell may be created by brain force through exercise, but mental and physical exercise always exhausts it for the time being at least, but lost energies are again restored by rest and inhalation. After violent exercise the brain, having expended its force, its cells are refilled by rapid respiration, and as soon as they are refilled the breathing becomes normal. If the physical exercise has been violent and prolonged, the brain cells become so exhausted that they cannot recuperate at once, therefore cannot supply the full flow of force required to keep up the strength, when the muscles become lax and tired. This fact is made plainer by the experience of a debauchee whose brain was under the influence of strong drink for the space of 24 hours, and during which time he is dreaming horid dreams and raving in delerium, thus exhausting the contents of his brain cells, and reducing his vital forces; but all of this time he is prone on his bed with all his muscles in a relaxed state, a condition necessary for rest and recuperation, and therefore he should become strong and active on his feet again as soon as he is sober enough to stand; but, on the contrary, his muscles are scarcely able to support his staggering body because his brain force has been exhausted by intense and violent thoughts, which requires hours for it to recuperate sufficiently to support the usual strength of the muscular system.

Nature completes the human machine at the moment of maturity, beyond which point of development man has no power to extend nor to change it except by violence. The body is ruled by the brain and not the brain ruled by the body. There exists no will-power, physical strength, nor mind independent of the brain. If the brain and internal organs could be reconstructed every seven years, as we have been instructed by very high authority that they can, then life would continue forever on the same principal as that of the old lady's socks that would last a lifetime, saying that she could knit socks that would last a lifetime by refooting them every winter and retoping the same every other winter.

If new brain cells are continually forming, their strength could not grow less, consequently they would support the current of vitality all the time, and, since the body is supported by the brain, it could not wear out. If the renewing process of the body is incessantly going on, there is no reason why life should not continue indefinitely; but it never does. We have all heard of a Methusala, the wandering Jew and tibitian monstrosity, but none of them have been caught and placed on exhibition to demonstrate their existence or to prove the presence of the inexorable forces in nature, which were necessary to produce them, or as freaks that live independently, or in defiance of natural laws; therefore it can safely be said that life is never prolonged beyond the extreme limit of ordinary existence. When one or more of the divisions of the brain fail in its functions, life goes out of the entire encephalon, be it a Jew, Gentile, or a Mahatma, and he who says that he can prolong life by any artificial means, beyond a reasonable limit. is not a profitable companion or councilor.

The brain is composed of layers, and all have their physical duties to perform, and when each has served its time in executing the physical functions of the body then life goes out. When once set in motion the brain never stops, except by violence, until some part of it is worn out, and when it does stop from natural causes, man has no power to infuse life into its wornout cells, for the atoms of matter, which were enlisted in the service of life at its beginning, have served out their time and cannot be induced to re-enlist, while other atoms are not permitted to take their places since the Zodiacal brain was created, prior to birth, in nature's secret laboratory and cannot be returned to the hands which created it, for repairs.

For the above reasons man grows old and infirm, woman becomes wrinkled and fades. The rose of youth departs from the cheek of the blushing maid, the luster from the sparkling eyes. The step of pride loses its firmness, the mind becomes enfeebled, memory fleas away, the animated machine refuses to continue its labors and stillness reins supreme, in the once active citadel of life.

THE HOROSCOPE OF JUSTICE RATHBONE.

The Founder of the Order of the K. of P. was born at the rising of the Sun, Jupiter and Mercury in the sign Scorpio, which gave him a large head, active brain and force of character. Scorpio, rising, gave him a large, stout body, endowed with strength and activity. The Sun and Mercury, rising, gave him force of character, magnetic power and social influence. Mercury and the Sun, rising in the sign Scorpio, made him studious, secretive and thoughtful, which are prerequisite to the successful practise of Medicine, for which Mr. Rathbone would have been especially adapted. Jupiter, joined to the Sun, gave him a strong social development. Venus, in the house of friends, increased that development; consequently, his mind led him into the social channels, which caused him to study man's social and fraternal relations to each other; but this desire alone would not have accomplished Mr. Rathbone's object of brotherhood.

He was born, like Lincoln and Darwin, when Mercury and Uranus were receiving a trine influence from each other, which gave him originality of thought, independence in action and a large degree of intuition. This magnetic strength, coupled with his mental acuteness, gave him the power to thrill the fraternal world while rehearsing the products of this master mind, and made his name famous the world over.

HOROSCOPE OF THOMAS A. EDISON.

T. A. Edison was born Feb. 11, 1846, one year earlier than the date given for his birth. This I know to be a truth from the fact that Mr. Edison could not be the genius he is if born in '47; but Feb. 11, '46, describes him as he is.

He was born at 3.30 p.m., at which time Scorpio was rising, and he is therefor ruled by Mars, which I find in sign Taurus, joined to Jupiter, their influenc makes him self-confident, self-willed, firm and determined. He aspires to leadership and takes quiet pride in being famous.

The Moon, at Mr. Edison's birth, was in the midheaven, which has a tendency to give popularity and public notority. The Moon, Mars and Jupiter, in trine to each other, also gives him force, energy and a degree of popularity, the influence and support of which is to give fame; but they also show a large, strong, active brain. I further find Mercury in sextile to Uranus, which gave to Mr. Edison originality of

thought, independence of action and an inquisitive
and inventive turn of mind.

Mr. Edison was not made for a plodder; he is a
natural investigator.

He takes nothing for granted;he likes to wrestle with
the unknown and unseen things of nature. Had Mr.
Edison been born 50 years sooner, before electricity
was so well known, he doubtless would have studied
chemistry or some mechanical inventions. Like Darwin and Lincoln he was born when Mercury and
Uranus were in friendly aspect to each other. Uranus
gives scientific and inventive ability; but, in other
particulars, the strength of the horoscope determines
his bent of mind.

I use the foregoing horoscope to show that a
benefic influence of Mercury to Uranus always gives
a mind that will branch out in a new and strange
field of operation or investigation, because when the
brain is large and active it always produces something
new and strange, if not startling. The size and condition of the brain can be known by the planets.

There are many persons born, when Mercury and
Uranus are in friendly aspect to each other, that accomplish nothing above the ordinary, because their
brains are small and inactive.